celebrating mathematical mistakes

How to Use Students' Thinking to Unlock Understanding

Nicole M. Wessman-Enzinger & Natasha E. Gerstenschlager

Solution Tree | Press

Copyright © 2025 by Solution Tree Press

Materials appearing here are copyrighted. With one exception, all rights are reserved. Readers may reproduce only those pages marked "Reproducible." Otherwise, no part of this book may be reproduced or transmitted in any form or by any means (electronic, photocopying, recording, or otherwise) without prior written permission of the publisher.

555 North Morton Street
Bloomington, IN 47404
800.733.6786 (toll free) / 812.336.7700
FAX: 812.336.7790

email: info@SolutionTree.com
SolutionTree.com

Visit **go.SolutionTree.com/mathematics** to download the free reproducibles in this book.

Printed in the United States of America

Library of Congress Cataloging-in-Publication Data

Names: Wessman-Enzinger, Nicole M., author. | Gerstenschlager, Natasha E., author.
Title: Celebrating mathematical mistakes : how to use students' thinking to unlock understanding / Nicole M. Wessman-Enzinger, Natasha E. Gerstenschlager.
Description: Bloomington, IN : Solution Tree Press, [2025] | Series: Growing the mathematician in every student | Includes bibliographical references and index.
Identifiers: LCCN 2024016209 (print) | LCCN 2024016210 (ebook) | ISBN 9781960574329 (paperback) | ISBN 9781960574336 (ebook)
Subjects: LCSH: Mathematics--Study and teaching.
Classification: LCC QA11.2 .W437 2025 (print) | LCC QA11.2 (ebook) | DDC 510.71--dc23/eng20240802
LC record available at https://lccn.loc.gov/2024016209
LC ebook record available at https://lccn.loc.gov/2024016210

Solution Tree
Jeffrey C. Jones, CEO
Edmund M. Ackerman, President

Solution Tree Press
President and Publisher: Douglas M. Rife
Associate Publishers: Todd Brakke and Kendra Slayton
Editorial Director: Laurel Hecker
Art Director: Rian Anderson
Copy Chief: Jessi Finn
Senior Production Editor: Sarah Foster
Copy Editor: Jessi Finn
Text and Cover Designer: Julie Csizmadia
Acquisitions Editors: Carol Collins and Hilary Goff
Content Development Specialist: Amy Rubenstein
Associate Editors: Sarah Ludwig and Elijah Oates
Editorial Assistant: Anne Marie Watkins

GROWING THE MATHEMATICIAN IN EVERY STUDENT COLLECTION

Consulting Editors: Cathy L. Seeley and Jennifer M. Bay-Williams

No student should feel they're "just not good at math" or "can't do math"!

Growing the Mathematician in Every Student is a collection of books that brings a joyful positivity to a wide range of topics in mathematics learning and teaching. Written by leading educators who believe that every student can become a mathematical thinker and doer, the collection showcases effective teaching practices that have been shown to promote students' growth across a blend of proficiencies, including conceptual development, computational fluency, problem-solving skills, and mathematical thinking. These engaging books offer preK–12 teachers and those who support them inspiration as well as accessible, on-the-ground strategies that bridge theory and research to the classroom.

Consulting Editors

Cathy L. Seeley, PhD, has been a teacher, a district mathematics coordinator, and a state mathematics director for Texas public schools, with a lifelong commitment to helping every student become a mathematical thinker and problem solver. From 1999 to 2001, she taught in Burkina Faso as a Peace Corps volunteer. Upon her return to the United States, she served as president of the National Council of Teachers of Mathematics (NCTM) from 2004 to 2006 before going back to her position as senior fellow for the Dana Center at The University of Texas. Her books include *Faster Isn't Smarter* and its partner volume, *Smarter Than We Think*, as well as two short books copublished by ASCD, NCTM, and NCSM: (1) *Making Sense of Math* and (2) *Building a Math-Positive Culture*. Cathy is a consulting author for McGraw Hill's *Reveal Math* secondary textbook series.

Jennifer M. Bay-Williams, PhD, a professor at the University of Louisville since 2006, teaches courses related to mathematics instruction and frequently works in elementary schools to support mathematics teaching. Prior to arriving at the University of Louisville, she taught in Kansas, Missouri, and Peru. A prolific author, popular speaker, and internationally respected mathematics educator, Jenny has focused her work on ways to ensure every student understands mathematics and develops a positive mathematics identity. Her books on fluency and on mathematics coaching are bestsellers, as is her textbook *Elementary and Middle School Mathematics: Teaching Developmentally*. Highlights of her service contributions over the past twenty years include serving as president of the Association of Mathematics Teacher Education, serving on the board of directors for the National Council of Teachers of Mathematics and TODOS: Mathematics for ALL, and serving on the education advisory board for Mathkind Global.

Acknowledgments

Solution Tree Press would like to thank the following reviewers:

Lindsey Bingley
Literacy and Numeracy Lead
Foothills Academy Society
Calgary, Alberta, Canada

Kelly Hilliard
GATE Mathematics Instructor
 NBCT
Darrell C. Swope Middle School
Reno, Nevada

Shanna Martin
Middle School Teacher &
 Instructional Coach
School District of Lomira
Lomira, Wisconsin

Paula Mathews
STEM Instructional Coach
Dripping Springs ISD
Dripping Springs, Texas

Demetra Mylonas
Education Researcher
Headwater Learning Foundation
Calgary, Alberta, Canada

Janet Nuzzie
District Intervention Specialist
Pasadena ISD
Pasadena, Texas

Rachel Swearengin
Fifth-Grade Teacher
Manchester Park Elementary
Lenexa, Kansas

Visit **go.SolutionTree.com/mathematics**
to download the free reproducibles in this book.

Table of Contents

Reproducibles are in italics.

ABOUT THE AUTHORS xv

INTRODUCTION 1

PART I: Celebrating Mathematical Mistakes — 7

CHAPTER 1 — 11
A Shifted View of Mistakes

Shifting Views of Mistakes 12
Celebrating Mathematical Mistakes 15
Concluding Remarks 17
Reflection Questions 17
Chapter 1 Application Guide 18

CHAPTER 2 — 21
Beautiful and Powerful Mistakes

Beauty and Power in Mathematics 22
Negative Integers as a Creative Space for Mistake Making 25
Concluding Remarks 34
Reflection Questions 35
Chapter 2 Application Guide 36

CHAPTER 3 — 39
Factual, Procedural, and Conceptual Mistakes

- Types of Mistakes 40
- Concluding Remarks 58
- Reflection Questions 58
- Chapter 3 Application Guide 60

CHAPTER 4 — 63
Mistakes by Mathematicians

- A Continuum of Favorite Mistakes 65
- How to Elicit and Celebrate Mistakes 71
- An Act of Celebrating 73
- Concluding Remarks 74
- Reflection Questions 74
- Chapter 4 Application Guide 76

PART II: Responding to Mathematical Mistakes in Action — 79

CHAPTER 5 — 83
Two Foundational Instructional Strategies for Examining Mistakes

- Laying the Foundation for an Asset-Based Perspective 85
- Using the Unknown Student Work Strategy 88
- Selecting Tasks to Promote the Inspection-Worthy Mistakes Strategy 94
- Converting or Revising Existing Tasks 97
- Concluding Remarks 99
- Reflection Questions 99
- Chapter 5 Application Guide 100
- Chapter 5: Two Foundational Instructional Strategies for Examining Mistakes 101

Table of Contents

CHAPTER 6 — 103
Changing Minds in Mathematics

- Changing Minds 104
- Concluding Remarks 118
- Reflection Questions 119
- *Chapter 6 Application Guide* 120
- *Chapter 6: Changing Minds Task Structure* 121

CHAPTER 7 — 123
This or That Tasks

- This or That Task Structure 124
- How to Create This or That Tasks 125
- Concluding Remarks 130
- Reflection Questions 131
- *Chapter 7 Application Guide* 132
- *Chapter 7: This or That Task Structure* 133

CHAPTER 8 — 135
Invented Notation and Language

- Invented Notation and Language 137
- How to Use Invented Notation and Support Invented Language in the Classroom 147
- Concluding Remarks 149
- Reflection Questions 150
- *Chapter 8 Application Guide* 151
- *Chapter 8: Invented Notation and Language* 152

CHAPTER 9 — 155
Mathematical Games

- Mathematical Play 156
- Gameplay 161
- Concluding Remarks 169
- Reflection Questions 169
- *Chapter 9 Application Guide* 170
- *Chapter 9: Resources for Mathematical Gameplay* 171

CHAPTER 10 — 175

Mistakes in Action

Revisiting the Big Ideas From Each Chapter 177

Concluding Remarks 185

Reflection Questions 185

Chapter 10 Application Guide 186

EPILOGUE — 189

REFERENCES AND RESOURCES — 193

INDEX — 201

About the Authors

**Nicole M. Wessman-Enzinger, PhD,** is an associate professor of education at George Fox University in Newberg, Oregon. Her main role at the university is preparing future teachers (K–12) to teach mathematics, but her favorite part of this work is helping her future teachers redefine their relationships with mathematics. She enjoys exploring mathematics deeply and creatively with them. Formerly, she was a high school mathematics teacher in Illinois, where she learned from her students and her colleagues (like the incredible mathematics teachers at the Metropolitan Mathematics Club of Chicago).

Nicole is a member of various professional organizations, including the International Group for the Psychology of Mathematics Education and the National Council of Teachers of Mathematics. She regularly presents at national and international mathematics education conferences, and her work on students' conceptions of numbers, mathematical mistakes, and joy in mathematics is published in various journals. She most enjoys studying children's thinking about numbers as a way of recognizing students as mathematicians in classrooms.

Nicole received a bachelor's degree in mathematics and mathematics education from Olivet Nazarene University (Bourbonnais, Illinois), a master's degree in mathematics and mathematics education from DePaul University (Chicago, Illinois), and a PhD from Illinois State University (Normal, Illinois).

The best part of Nicole's day is supporting her own young mathematician, Mathilda, in experiencing the beauty and joy in mathematics.

Natasha E. Gerstenschlager, PhD, is a senior learning scientist in mathematics for Houghton Mifflin Harcourt. Formerly, she was an associate professor of mathematics education in Kentucky, where she taught mathematics courses for current and prospective teachers and developed and led professional development sessions on innovative, research-based teaching practices.

Natasha is a member of the American Educational Research Association and the National Council of Teachers of Mathematics, for which she serves as a department editor. She has presented on her work at national and international conferences, and her ideas for improving mathematics teaching across grade levels have been published in various journals. She enjoys connecting the ideas in research to the application of practice for mathematics educators.

In addition to her work, and more importantly, Natasha delights in helping her two children experience mathematics in their daily lives (particularly during camping adventures and board games) so that they can develop a love for and a joy in mathematics that go beyond the classroom.

Natasha received a bachelor's degree and a master's degree in mathematics and a PhD in mathematics and science education, all from Middle Tennessee State University.

To book Nicole M. Wessman-Enzinger or Natasha E. Gerstenschlager for professional development, contact pd@SolutionTree.com.

Introduction

What is one thing that is guaranteed to happen in mathematics and science for every learner? Mistakes. Every one of your students will make a mistake when they are learning—whether that mistake is seen by all or they neatly tuck it away or internally muse about it during class. What sets apart those who eventually become mathematicians and scientists is they are not afraid to make mistakes. As astrophysicist and educator Neil deGrasse Tyson says, "I love being wrong 'cause that means in that instant, I learned something new that day" (Sagal, 2015).

In other words, mathematicians and scientists recognize mistake making as an opportunity for learning. In fact, we know from cognitive science that when people make a mistake, new synapses form in their brain (Boaler, 2016), which doesn't happen when they do not make a mistake. How cool is that? So, this has caused us to wonder, "Why, with all these amazing mathematical and scientific minds celebrating mistakes, and this research showing mistakes' value, do students and teachers still fear making mistakes on their mathematical journeys?" As K–12 educators, we have seen how viscerally our mathematics students react to making a mistake; many sweat, get anxious, or even cry. This is not how mathematical learning should be. We think that deficit-based views of mathematical mistakes are at the core of bad feelings about and bad experiences with mathematics. When we decided to write this book, our goal was to celebrate mathematical mistakes, countering the experiences many students and teachers have had with mathematics.

During our years spent teaching mathematics, we have gathered a variety of experiences in helping our students understand the value of mistakes. Not everything we have tried through these experiences has helped our students see how mistakes are celebratory and beautifully necessary. In this journey, we have made our own mistakes, but from those, we have learned a lot about how best to encourage and support mathematical mistakes. We decided to compile these experiences and the tools we have used successfully into this practitioner-based book for our fellow K–12 educators. Furthermore, we want you to have opportunities to learn and reflect on how beautiful, powerful, and celebratory mathematical mistakes are. We hope you celebrate mathematical mistakes with us. And we hope our book adds practical applications to your tool kit so you can help students in this same endeavor.

This book will be helpful for anyone teaching mathematics (for example, K–12 teachers, mathematics coaches, and paraprofessionals). We envision that it will best serve teams of mathematics educators in grade-level cohorts, vertical planning groups, or school-based teams in a book study or professional learning setting. We hope all who read this book walk away with ways to develop productive and healthy mistake making in their mathematics classrooms, and they adopt the perspective that mistakes are necessary, beautiful, powerful, and essential for learning.

We have organized the book into two parts. Part 1, "Celebrating Mathematical Mistakes," introduces reasons why mistakes are powerful, different types of mistakes that can be made in a mathematics classroom, and examples of mistakes from mathematicians (both famous mistakes and mistakes made by everyday mathematical practitioners). This part provides the foundation for why mistake making is important, beautiful, and powerful. We envision this part as providing the groundwork for the second half of the book.

Part 2, "Responding to Mathematical Mistakes in Action," presents the reader with practical tools for productively developing mathematical mistakes in the classroom. Here, we provide examples across K–12 and across content areas to demonstrate how these ideas are not specific to elementary students or to one specific subdomain of mathematics. These ideas can even be used in postsecondary education! We envision readers taking each chapter in part 2 as an opportunity to try something new in the classroom.

Part 1 (chapters 1–4) develops the readers' appreciation and understanding. In chapter 1, we reconceptualize mistakes for readers—that is, as opportunities for learning and trying again instead of dead-ending. We define deficit-based perspectives versus asset-based perspectives as they relate to mathematical mistake making. We share why mistake making is a human activity to celebrate rather than avoid.

In chapter 2, we introduce the idea that mistakes are beautiful and powerful. We define what we mean by *beauty* and *power* and dig deeper into mistake making's inherent presence in the mathematics field. We provide real examples of beautiful and powerful mistakes from our teaching experiences and explore what specifically about these mistakes makes them beautiful and powerful.

In chapter 3, we present the terms *factual mistakes*, *procedural mistakes*, and *conceptual mistakes*. We explore what these mean and what the mistakes look like in classrooms. We also unpack how each of these has a place in the learning trajectory, with some being especially powerful for developing a deep understanding of mathematics. To assist in this conversation, we provide vignettes based on our teaching experiences to illuminate the different types of mistakes and how they can be used.

In chapter 4, we share some mistakes made by mathematicians (both famous and everyday mathematicians and practitioners) to demonstrate the humanity within mistake making. We also use these mistakes to illuminate how acceptable and expected mistake making is in the field. People commonly think professional mathematicians don't make mistakes, but this is a misconception. This chapter debunks that myth.

Part 2 (chapters 5–10) develops readers' application and praxis. Praxis leads the discussion, and we encourage readers to envision how what we describe can look in their classrooms and schools. In chapter 5, we begin by presenting practical ways to implement the ideas presented in part 1. This chapter shares two easy-to-implement instructional strategies that set the foundation for powerful mistakes and allow for inspection of and respect for those mistakes. The first strategy involves using unknown student work as a means for examining and respecting mistakes, which sets the tone in the classroom and amplifies the idea that mistakes are useful and expected. The second strategy involves using carefully selected tasks that are known to elicit conceptual mistakes, which ensures those powerful mistakes that can move students along the learning trajectory happen organically.

In chapter 6, we highlight the Changing Minds Task Structure and how changing minds is a creative act and pedagogical tool that normalizes shifting one's thinking about a strategy or a solution toward which one has worked. By first seeing changing minds as a creative opportunity, teachers can then consider ways to implement strategies that encourage students to change their minds. Finally, we share how teachers can have students investigate another student's work in which they changed their mind while solving the task.

In chapter 7, we unpack the This or That Task Structure. As teachers, we often face situations where we want to explore mistakes, but we know that students must get to a correct strategy and solution for the sake of doing well on assessments. The This or That Task Structure allows us to honor correct and incorrect solutions at the same time, demonstrating the value of mistakes to deepen conceptual learning.

In chapter 8, we share supporting students' invented notation and language as a strategy for reinforcing creativity and agency in mathematics. Supporting invented notation and language will naturally result in mistake making. And interestingly, some things that seem like mistakes are not really mistakes at all.

In chapter 9, we highlight ways in which play and games offer students a safe environment in which to make mistakes and learn from those mistakes. We unpack the tenets of play and what play means within mathematics. We share ways in which you can incorporate play into your mathematics classroom. Then, we explain the purpose of games in conceptual understanding and how you can include games in the daily structure of mathematics without forsaking your pacing.

In chapter 10, we bring together the ideas from the previous chapters, describing how they all can play well in a mathematics lesson. We visit Mrs. W's classroom to see how she incorporates the ideas presented in the book, pausing during her instruction to highlight specific parts and bring your attention to important features.

For each chapter, we begin with a big idea that captures the central message and can serve as a guidepost while you read, and then we share a reflection from one of our personal histories that sets the foundation for the chapter. Throughout the chapters, we include Pause and Ponder questions at various stopping points. We envision you and your colleagues using these for personal journaling or discussion with others in a book-club environment. We hope these instigate further conversations for deeper learning.

Introduction

We also provide simple ways you can incorporate what you are learning into your classroom, signified with a light bulb icon to indicate passages where you consider how you can use mistakes to adjust instruction to drive learners toward the learning goal.

Each chapter ends with reflection questions and an application guide to help you review the chapter's content.

Finally, throughout the chapters, we highlight student work based on our experiences as teachers through vignettes and recreated student samples. We encourage our readers to begin a collection of their own students' work and mistakes so that they can learn from and leverage these gold mines of knowledge.

In closing, we hope that our readers walk away from this book with inspiration, joy, and hope that, as a field, mathematics education can move from correcting mistakes to expecting mistakes, from avoiding mistakes to unpacking mistakes, and from fearing mistakes to celebrating mistakes.

PART 1

Celebrating Mathematical Mistakes

In part 1 of this book (chapters 1–4, pages 11–76), we explore mistakes as things worth celebrating to excite and inspire you to use them in your classroom more. Valuing and using mistake making can help you leverage students' learning forward.

If we wish to support students as mathematicians, then we need to facilitate opportunities for creativity in mathematics. When you support students to be creative in mathematics (for example, to invent solutions or create patterns), students will naturally make mathematical mistakes. And we should celebrate mistakes. Students learn more from their mistakes than their successes (Boaler, 2016), and making mathematical mistakes is part of authentically doing mathematics (National Council of Teachers of Mathematics [NCTM], 2014; Smith & Stein, 1998). This part sets the stage for celebrating mathematical mistakes, addressing the following three main ideas.

1. Understanding why you should honor mistakes
2. Celebrating mathematical mistakes
3. Envisioning how you can celebrate mistakes in your classroom

Welcome to the celebration of mistakes, where we all are mathematicians!

CHAPTER 1

A Shifted View of Mistakes

BIG IDEA

Developing mathematicians means helping students shift from deficit-based views to asset-based views of mathematical mistakes.

> Before reading this chapter, take a moment to reflect on or journal about the following questions: Do you value mathematical mistakes? If so, in what ways do you value them?

A REFLECTION FROM NATASHA

My K–12 experience in mathematics instilled in me that I should avoid making mistakes, and if I were to be any good at mathematics, I had to be excellent at minimizing my mistakes. Naturally, when I began my studies as a mathematics major in college, I focused on making minimal mistakes in my work, and I thought my mathematics professors never made mistakes. Although some of my professors held a deficit-based view of mistakes, one professor encouraged my classmates and me to write our work in ink. *Gasp!* We all had our beloved pencils and erasers because we *had* to hide our mistakes by erasing them. He said when (not if) we made a mistake, we should simply strike it out cleanly and continue with our work. When he made a mistake in a lecture, he would ask us to inspect the mistake by posing questions such as, "Where do you think I went wrong? What did I do right? How could we use this to move forward?" He also took the same approach when students presented their work. By having us write our work in ink, he made us realize our mistakes were not to be hidden with an eraser but to be expected, analyzed, and documented for learning. For the first time, I saw myself, someone who regularly made mistakes, as a mathematician.

★ ★ ★

As you read this chapter, I hope you reflect on an experience that shifted your perspective on mistakes in the mathematics classroom from a deficit-based to an asset-based one as we discuss shifting views of mathematical mistakes and celebrating them.

Shifting Views of Mistakes

A German proverb, "Aus Schaden wird man klug," translates to "Failure makes smart" (GermanPod101, 2021). Learning happens alongside being wrong, failing, and making mistakes. A problem of practice for most mathematics teachers at all levels is that their students often view mathematical mistakes negatively (Leighton, Guo, & Tang, 2021; Yildiz, 2013), which can cause anxiety or fear about mathematics (Di Martino & Zan, 2013; Ganley, Schoen, LaVenia, & Tazaz, 2019). This fear of making mistakes in mathematics inhibits the learning experience. Therefore, reducing anxiety within mathematical experiences could enhance thinking and learning about mathematics (Buckley et al., 2021; Stoehr & Olson, 2023). How do we shift views about mathematical mistakes so that our students can have better mathematical experiences?

PAUSE AND PONDER

How do you define a mathematical mistake?

_____
_____
_____
_____
_____
_____
_____
_____
_____
_____

Shifting from deficit-based to asset-based views of mathematical mistakes can help students develop into fearless mathematicians. To encourage this, teachers must focus on developing students' mathematical identities. Mathematical identities are "the dispositions and deeply held beliefs that students develop about their ability to participate and perform effectively in mathematical contexts and to use mathematics in powerful ways across the contexts of their lives" (Aguirre, Mayfield-Ingram, & Martin, 2013, p. 14). One way teachers can develop these mathematical identities is by "vigilantly viewing student attributes as assets rather than deficits" (Allen & Schnell, 2016, p. 401). The disposition of a fearless mathematician entails making mistakes and learning from mistakes. By approaching mistakes from an asset-based perspective, teachers and students view mistakes "as points of mathematical interest and conversation" (Allen & Schnell, 2016, p. 403), like mathematicians. When you support asset-based views of mistakes, you provide opportunities to counter the narrative that mistakes in math class are shameful or embarrassing. In fact, mathematics educators Kasi Allen and Kemble Schnell (2016) describe that:

> Only in classrooms where errors serve as a cause for inquiry, even celebration, can we cultivate the kind of social safety and alternative definitions of mathematical success that allow every student to truly contribute in meaningful ways that will be honored and respected by every other student in the room. (p. 403)

To make the shift to more asset-based views of mathematical mistakes means seeing mathematics as more than operations or procedures that help one obtain a solution (Su, 2020).

While your students are working on a task, pause them *before* they come to a solution. Students will have imperfect drawings, equations, and mathematical notations as they create strategies. We refer to these messy written artifacts that capture their mathematical thinking prior to their arrival at a solution as *strategy work*. Ask students to share a mistake they made during their strategy work that will help them get back on track. This technique begins the shift toward asset-based views of mistakes because students see their mistake as valuable in their journey toward a solution. Focusing on the strategy work centers the process rather than only the solution. This gives you opportunities to ask questions like, "What worked well for you? What did not?" Regardless of getting correct or incorrect solutions, students will pivot toward reflecting on what they did and why they did it.

Before we discuss mistakes, let's look closely at mathematics. Doing mathematics is recognizing the beauty of logic, structures, and argument. It is explaining and justifying reasoning (Bieda & Staples, 2020; Brown, 2017), critiquing the reasoning of others (National Governors Association Center for Best Practices [NGA] & Council of Chief State School Officers [CCSSO], 2010; Wagganer, 2015), and revising thinking (Jansen, 2020). From mathematicians, we know that mathematics is a time-intensive process of which mistakes are an inherent component. Mathematics is messy because a lot of it is creating, playing with, and revising ideas. Yes, mathematicians are in pursuit of a correct solution, a logical strategy, or an explanation; however, they do not study geometry, topology, or knot theory because they care about being correct. Rather, mathematicians care about finding beauty in mathematics, listening to and learning from others, solving important problems, striving toward efficiency and optimization, and creating structures that support change.

Now, let's consider the word *mistake*. Its prefix, *mis-*, means "badly" or "wrong" (Mis-, n.d.). Yet, if our view of mathematics solely focuses on correct versus incorrect solutions, then we fail to see the true nature of mathematics. Therefore, in this book, we playfully revised the word *mistake* to *mistake*. We have used this modified version in our interactions with each other and our students. By striking through the deficit prefix *mis-*, we are explicitly countering the deficit connotations of *mistake* to inspire you and your students to shift your views about mathematical mistakes. Although we use *mistake* throughout this book so as not to be confusing, we hope that when you see the words *mathematical mistakes*, you read them as the asset-based versions.

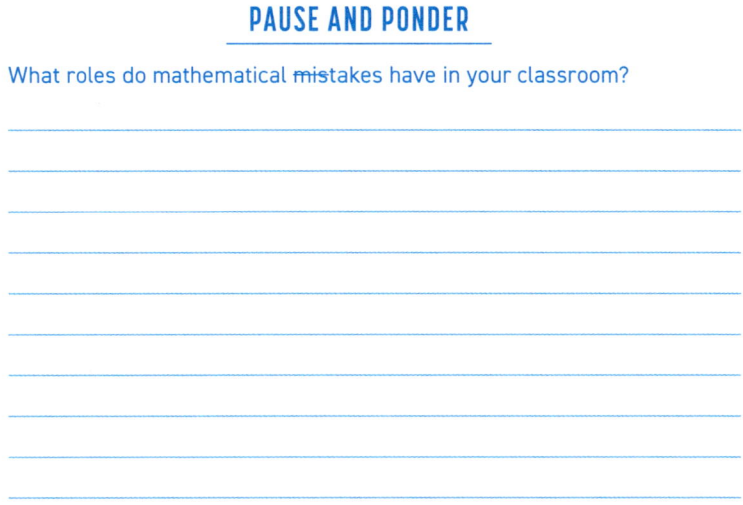

PAUSE AND PONDER

What roles do mathematical ~~mis~~takes have in your classroom?

_____
_____
_____
_____
_____
_____
_____
_____
_____
_____

On the journey of shifting from mistakes to ~~mis~~takes for more asset-based views, we have come to the following ways of thinking about mistakes.

- A mistake is an incorrect solution or an in-progress strategy that enhances mathematical learning.
- Mistakes are attempts at a problem that bring you closer to solving the problem.
- Mistakes are ways you previously saw yourself (or others) within mathematics that you are changing.

Celebrating Mathematical Mistakes

When you type "mistake making in mathematics" into the Google search engine, it suggests the most searched-for related entries include "how to avoid making mistakes in mathematics" and "how to stop making mistakes in mathematics." Why would people want to stop or avoid making mistakes in mathematics? Instead, they should celebrate mathematical mistakes because these are at the heart of grappling with a problem, inventing strategies, and exploring a curiosity. Without mistakes, people would be unable to fully do mathematics. Thus, we need to normalize mathematical mistakes by celebrating them rather than avoiding them.

> At the beginning of the year, when you are establishing classroom norms, it may be useful to create a celebration board where students can post their mathematical mistakes. Have each student use a sticky note to keep track of their mistakes during their mathematics lesson and post it (anonymously if they wish) after class. At the end of the week, select a few mistakes to share and celebrate to let students see that you value, expect, and encourage their mathematical mistakes. The intent of this activity is to celebrate the mistake; although you're not intended to discuss the strategy at this moment, you could decide to do that after the sharing.

Furthermore, celebrating mathematical mistakes humanizes mathematics since mistakes are an inherent part of life. Therefore, mistakes are part of mathematics too. In fact, mathematical mistakes should be "expected, inspected, and respected" (Seeley, 2016, p. 26) in the classroom.

PAUSE AND PONDER

Think about mathematics educator Cathy L. Seeley's (2016) statement that mathematical mistakes should be "expected, inspected, and respected" (p. 26). What does it look like in practice to expect mistakes? To inspect mistakes? To respect mistakes?

Embracing mathematical mistakes in the classroom is crucial for supporting students to develop in four ways: (1) productive struggle (NCTM,

2014; O'Dell, 2018), (2) growth mindsets (Boaler, 2016; Dweck, 2006; Dweck & Yeager, 2019; Sun, 2018), (3) identities as mathematicians (Aguirre et al., 2013), and (4) conceptual understanding (Kilpatrick, Swafford, & Findell, 2001). Engaging in productive struggle requires making mistakes, and productive struggle is related to joy in mathematics (O'Dell, 2018) and growth mindsets (Townsend, Slavit, & McDuffie, 2018). Mistakes can be joyful, and when there is joy in mathematics, students may play with mathematical content in the same ways as mathematicians (Parks, 2020). Therefore, mistakes are worth positioning at the forefront of the mathematics classroom as worthy and important for learning. An important way to support this positioning of mathematical mistakes is to intentionally celebrate them.

Concluding Remarks

In this chapter, we encouraged you to shift from deficit-based views to asset-based views of mathematical mistakes. We established this inaugural chapter in a hopeful nature of leaning into change and growth in how you see yourself and your students as doers of mathematics. If research mathematicians can make mistakes as part of their practice, then why should we not all make mistakes? In this book, we aim to debunk the myth that students should avoid mistakes in mathematics. Instead, we offer a counterview that embraces and values mistakes. Celebrating mathematical mistakes will help you support students in growing and becoming mathematicians.

Reflection Questions

Use the following questions for reflecting on the ideas in chapter 1.

1. In what ways can you shift your thinking about mathematical mistakes to an asset-based perspective?

2. What are some ways you can shift your thinking from that of mistakes to mistakes?

3. Describe spaces in your own teaching where mathematics is more than correct and incorrect solutions. Where do mathematical mistakes show up in those spaces? How are they helpful?

4. How can you celebrate mathematical mistakes in your classroom?

Chapter 1 Application Guide

In this chapter, we redefined mathematical mistakes as ~~mis~~takes as a way to honor a shift to asset-based views of mistakes. Use the following application guide to connect the chapter's themes to your classroom.

Chapter Theme	Connection and Application to Your Practice
Having asset-based views of mathematical mistakes	Look for the good in students' mathematical mistakes. Is the mistake a nearly completed strategy? Is the mistake something that will enhance the students' thinking and learning?
Redefining mistakes as ~~mis~~takes	Playfully redefine the word *mistakes*, which can be written as ~~mis~~takes. One way to think about mistakes is that there are no mathematical mistakes; rather, there are only ways of thinking and conceptions. When you think of all students' thinking as conceptions, you start to approach supporting your students differently.
Celebrating mistakes	Honor students' mistakes and celebrate them in any way possible. Can you have a student present their work, even with a mistake, and highlight strengths? Honor the journey of mathematics, rather than the final destination of the solution, in your classroom.

Celebrating Mathematical Mistakes © 2025 Solution Tree Press • SolutionTree.com
Visit **go.SolutionTree.com/mathematics** to download this free reproducible.

CHAPTER 2

Beautiful and Powerful Mistakes

BIG IDEA

Defining beauty and power in mathematics requires teachers to think about how mathematical mistakes align with beauty and power. Seeing mistakes as beautiful and powerful mathematics provides opportunities for thinking about how to use mistakes in the classroom in different ways.

> Before reading this chapter, take a moment to reflect on or journal about the following questions: What three words do you associate with the phrase *mathematical mistakes*? Write them down, and reflect on them as you read this chapter.

A REFLECTION FROM NICOLE

I began my career as a public high school teacher and later became, and remain, a university professor who talks about and teaches mathematics education every day. As other educators do, I always consider myself a student as well.

During my education, it was wild to me how many times I heard my teachers and mathematics professors say to their students, "Mathematics is beautiful" or "This is powerful." Who said it wasn't beautiful or powerful? Who didn't find mathematics beautiful and powerful? I couldn't comprehend that. When illustrating this beauty, my teachers and mathematics professors showed Fibonacci numbers in the spirals of sunflowers, explained snowflakes and symmetry, and explored paradoxes (for example, Gabriel's horn has infinite volume but finite surface area). Unfortunately, not all my classmates were excited about mathematics and its beauty. As a teacher, I learned why students did not experience the beauty and power of mathematics.

When I became a mathematics teacher, I met many students who did not experience mathematical beauty the way that I did. And yet, the unique ways they thought and the diverse perspectives they shared on solving problems were more beautiful than the symmetry of a snowflake. I saw beautiful and powerful thinking about mathematics, even in my students' mistakes. In this chapter, Natasha and I offer a unique perspective on seeing mistakes as beautiful and powerful mathematics. We think that this is valuable to use in your classroom because when you see beauty and power in your students' mathematical mistakes, they will begin to see them as well. When students see their own beauty and power in mathematics, they will grow in their mathematical identities and mathematics. We also include vignettes that show the beauty and power in students' mathematical mistakes.

★ ★ ★

Beauty and Power in Mathematics

Imagine asking your students if the following statement is true: $0.9999\ldots = 1$. How would your students respond? We have asked this question of students in fifth grade, middle school, high school, and university. In our experience, nearly all students, or sometimes an entire class, answer it incorrectly. Most students say that $0.9999\ldots$ cannot equal 1. Indeed, it is mind-blowing that $0.9999\ldots = 1$. However, it is an amazing question to ask students because it sparks spicy debates when we ask them to justify their mistakes. Students' mistakes usually center on rounding $0.9999\ldots$ to 1, so we ask them if a number can be rounded if $0.9999\ldots$ is equivalent to 1. Typically, students base their mistakes on the argument that there is some finite distance between $0.9999\ldots$ and 1, like 0.0001 or 0.000000001. We counter that a finite distance between $0.9999\ldots$

and 1 assumes that 0.9999. . . has a finite number of digits, like 0.99, 0.999, or 0.9999999. Can they tell us exactly what number is between 0.9999. . . and 1, or the distance between 0.9999. . . and 1? The debates about whether 0.9999. . . = 1 are delightfully energetic and often full of "incorrect" mathematics, no matter the grade level. It is surprising to students (at all levels) that 0.9999. . . = 1.

Participating in mathematical argumentation and justifying mathematical conjectures can help develop students' conceptual understanding of mathematics (Rumsey & Langrall, 2016), which is especially great for surprising problems like this one. It is beautiful to watch our students grapple with the complexities of the infinite. Some of our younger students (fifth graders and middle schoolers) have eventually made the connection that ⅓ = 0.3333. . . and extended this reasoning to ⅗ or 1. One argument includes that if ⅓ = 0.3333. . . and 3 × ⅓ = 1, then 3 × 0.3333. . . = 0.9999. . . but this is a rich task that can be used at various levels. For example, our calculus students made connections to geometric series or even prepared algebraic proofs (for example, using if $x = 0.9999...$, then $10x = 9.9999...$ to create a proof). When students make connections within their mathematics and adjust their thinking, they get closer to correct solutions. But more importantly, their thinking becomes more powerful.

Of course, beauty and power in mathematics are subjective. What you see as beautiful or powerful about mathematics may not match what we see. Yet there are well-established attributes of mathematical beauty, as we present next. Then, we will define power in mathematics. After that discussion, we devote the bulk of this chapter to presenting a classroom vignette about negative integers through the perspective of five students.

DEFINING BEAUTY IN MATHEMATICS

Mathematicians have long revered the beauty of mathematics. In this chapter, we focus on three attributes of mathematical beauty.

1. Generating new ideas (Poincaré, 1910)
2. Recognizing both order and limitations within mathematics (Aristotle, 350 BC)
3. Eliciting surprise (Sinclair, 2004)

For example, in many ways, 0.9999. . . = 1 is surprising, which makes it beautiful. We have yet to encounter a mathematics class that is not surprised by this. And this surprise in mathematics leads to memorable experiences and deep discussions. The surprise also leads to wonder and curiosity, and becomes beautiful for that as well.

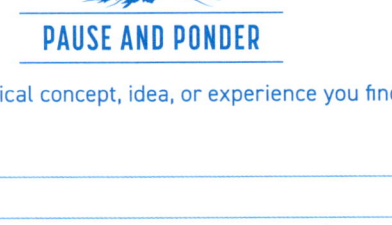

PAUSE AND PONDER

What is a mathematical concept, idea, or experience you find beautiful? Why is it beautiful?

___
___
___
___
___
___
___
___
___
___
___
___

DEFINING POWER IN MATHEMATICS

Mathematics professor Francis Su (2020) discusses creative power in mathematics as *sense making* and *structure identification*, which are mathematical tools that help all mathematics learners (from young students to career mathematicians) expand their current mathematics. Similarly, the Mathematical Practices in the Common Core State Standards include recommendations for sense making and use of structure (NGA & CCSSO, 2010). Mathematical Practice 1 of the Common Core State Standards is, "Make sense of problems and persevere in solving them" (NGA & CCSSO, 2010). Sense making as a powerful mathematical tool includes, but is not limited to, thinking about parameters, forming goals, making analogies, developing flexibility, and tackling challenges. A way of making sense of mathematics appears in Mathematical Practice 7, which suggests students "look for and make use of structure" (NGA & CCSSO, 2010). Students of all levels can look for structure and leverage that structure toward meaning making in mathematics.

Although we emphasize these well-established attributes of mathematical beauty and power within this chapter, we also share how we personally see beauty in mistakes and students' mathematics. Additionally, our perspective on beautiful and powerful mistakes will illuminate opportunities for growth in students' mathematical learning.

Negative Integers as a Creative Space for Mistake Making

Negative integers provide a space for students to engage in intellectual play in elementary school (Bofferding, Aqazade, & Farmer, 2018; Featherstone, 2000; Wessman-Enzinger, 2018)—particularly if students play with negative integers prior to formal instruction with them, which typically occurs in middle school (NGA & CCSSO, 2010). When students explore a mathematical idea, like negative integers, for the first time, they may not obtain correct solutions, or their justifications for correct solutions may not be robust. Such responses are often referred to as *mistakes*, but we know they are ~~mistakes~~. These ~~mistakes~~—students' early intuitions about operations with mathematical ideas—present us with opportunities to build on their thinking, like taking them on an exciting mathematical journey into negative integers. Mathematicians describe playing with mathematics and drawing on intuitions as part of the work they do that leads to valuing the beauty of mathematics (Sinclair, 2004).

To elicit mistakes that can be beautiful and powerful, allow students time to play with a new concept (for example, infinity) as they begin learning it. Provide them with open-ended tasks, read a book about the concept, ask students, "What do you know about this concept?" or give students time to explore a new manipulative in tandem with an open-ended task.

Next, we will share vignettes of four grade 5 students' strategies for integer multiplication. Taylor, Gertrude, Miguel, and Warren worked individually on integer multiplication problems. In each case, the students were thinking about the problems for the first time, so they shared their initial thinking. The teacher could expect mistakes because these students were seeing negative integers for the first time and inventing strategies for integer multiplication expressions (for example, -2×3) prior to any formal instruction with negative integers. In every example case, we share the ways we see beauty and power in the students' mathematics, mistakes and all.

We begin by sharing how Taylor and Gertrude individually solved -2×3 and how they each invented a way to multiply negative integers. Both were told they could solve the problem any way they wanted, and they had a variety of tools available, including empty number lines, linking cubes, two-color chips, and markers. They were told they could use some, all, or none of the materials as they shared their initial thoughts about the value of –2 × 3.

TAYLOR'S STRATEGY

The following excerpt illustrates Taylor's response to –2 × 3 and the beautiful mathematical contributions it offers.

> *Taylor: Hmm . . . so zero?*
>
> *Teacher: Zero? Can you tell me how you got that?*
>
> *Taylor: Since negative 2 isn't a real number . . . it's a number; it's not in the number that doesn't have the negatives.*
>
> *Teacher: Hmm.*
>
> *Taylor: And 3 times 2 is 6, but since it's negative, it's going to be zero.*
>
> *Teacher: OK. And, it's going to be zero because negative 2 . . .*
>
> *Taylor: It's lower than upper.*

This excerpt reveals Taylor's strategy when they saw –2 × 3 for the first time; they told the teacher that the solution to –2 × 3 is 0. The teacher asked Taylor why the answer is 0. Taylor recognized that 2 × 3 = 6, but they argued that –2 × 3 = 0 because –2 is "isn't a real number" and that –2 is more "lower than upper."

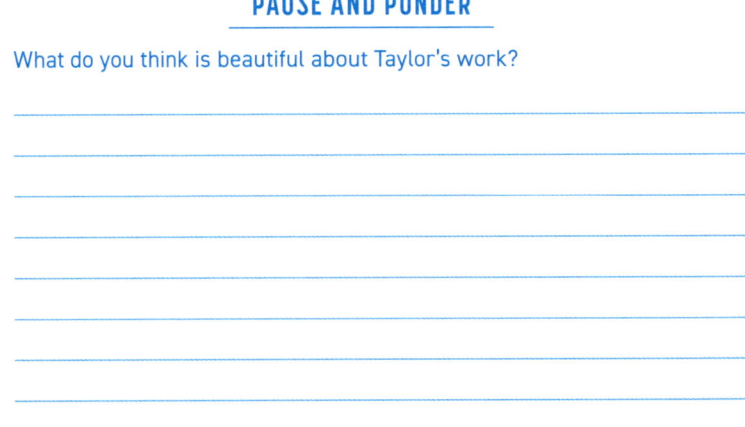

PAUSE AND PONDER

What do you think is beautiful about Taylor's work?

THE BEAUTY IN TAYLOR'S STRATEGY

Although Taylor stated an incorrect solution to –2 × 3, they *generated new ideas* around negative integers by creating a new strategy. Treating –2 differently than +2 is an important step in understanding both positive and negative numbers. For Taylor and many others, operating with a negative integer is challenging because –2 "isn't a real number" in the sense that it cannot be physically embodied in the way whole numbers can (Martínez, 2006). For example, 2 can be embodied by two objects (2 cookies), but this does not readily extend to negative numbers without adjustments (–2 cookies). Indeed, representing –2 has physical limitations. Taylor's *recognition of this limitation* led them to treat –2 as 0, and it is a beautiful limitation of negative numbers. In fact, mathematicians struggled with this same idea for centuries (Bishop et al., 2014), and Taylor's contribution honors this complexity.

Taylor's contribution of –2 × 3 = 0 is beautiful because it highlights the need to discuss the physical embodiment of negative integers. Yet Taylor was on the cusp of recognizing –2 × 3 may not be equivalent to 0, noticing –2 was in a different location than 0 or 2, as –2 is more "lower than upper." In engaging with negative integer multiplication, Taylor expressed the complex nature of negative numbers by highlighting the constraints in physical embodiment, an important first step for conceptualizing operations with negative numbers and much more.

GERTRUDE'S STRATEGY

The following excerpt highlights Gertrude's response to –2 × 3. We share this example to explore how her thinking could be enhanced by incorporating Taylor's beautiful mathematics although Gertrude gets the correct solution.

> *Gertrude, reaching for linking cubes: I'm going to pretend this is negative.* (Holds two white linking cubes.) *This is negative 2. A negative plus a negative would be a negative. So, if these are negatives, then that would be 3 times the 2 negatives, which equals 6 negatives.* (Writes –6.)

Gertrude counted out two negatives, physically representing them with two white linking cubes. Then, she counted out two more groups of two white cubes. This resulted in three groups of two white cubes (as shown in figure 2.1, page 28).

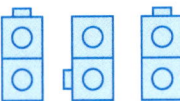

Figure 2.1: Linking cube representation of −2 × 3 = −6, where each linking cube represents −1.

Although this representation looks like 2 × 3 = 6, Gertrude mentally mapped a value of −1 to each cube. This is a novel construction that Gertrude invented. Her strategy differs from typical school instructional models for integer multiplication, such as jumping on a number line or using two sets of two-color chips.

THE BEAUTY IN GERTRUDE'S STRATEGY

For Gertrude's strategy, we highlight two attributes of mathematical beauty described earlier: (1) generating new ideas (Poincaré, 1910) and (2) eliciting surprise (Sinclair, 2004). Notably, Gertrude related the negative integers to something physical, the linking cubes. The constraints and challenges of physically embodying negative integers (Martínez, 2006) are often what students struggle with as they transition from whole numbers to negative integers. Yet Gertrude *created* a way to connect whole-number representations of multiplication to multiplication using negative integers, which is a beautiful physical representation. Another aspect of beauty we see is that Gertrude used the commutative property to her advantage. One interpretation of −2 × 3 could be −2 groups of 3 objects. Instead, Gertrude explicitly used the "groups of" interpretation of multiplication flexibly. She can see 3 × −2 as 3 groups of −2 and use the commutative property in her representation.

Not only did Gertrude generate a new idea, but we find that idea surprising. It surprises us because during her first time operating with negative numbers, Gertrude created a model that is not traditionally used in school mathematics (using linking cubes, instead of chips or a number line, for negative number operations) to help her obtain the correct solution.

For generating opportunities for beautiful and powerful mistakes, consider providing students with a manipulative tool kit for their group or table, but do not tell them which manipulative to use for the task at hand. This encourages creative and flexible thinking, and it provides fertile ground for innovative mistakes.

BETTER TOGETHER WITH BOTH CORRECT AND INCORRECT SOLUTIONS

Whether correct or incorrect, Taylor and Gertrude both thought about the physical embodiment of negative integers through negative integer multiplication. And they both offered glimpses into the beauty of mathematics. For example, they both generated mathematical ideas for making mental and physical adjustments in thinking about multiplication with negative numbers, which is ultimately why Taylor concluded that $-2 \times 3 = 0$, treating -2 as 0. However, Gertrude provided an example of an invented strategy for negative integer multiplication that supports a physical representation for $-2 \times 3 = -6$.

Taylor, who thought -2 could be equated to 0, could benefit from Gertrude's idea that -2 may be represented with two linking cubes. Additionally, Gertrude could benefit from Taylor's idea that -2 is difficult to physically represent—two linking cubes can represent both 2 and -2. Gertrude could also benefit from Taylor's idea of "lower and upper," as hearing that language may spark thoughts about representing multiplication on a number line rather than with discrete objects. Although Gertrude's representation of -2×3 is correct and works with discrete blocks, this type of representation alone, without a number line or reflection on the challenging nature of negative integers, will break down for -2×-3. So the discussion ought not end there. The beauty in Taylor's incorrect solution is that it highlights the need to think about negative numbers as relative numbers and consider additional models. Taylor's incorrect solution offers space for deepening meaning making about integers.

We have explored how Gertrude and Taylor thought about and solved the expression -2×3 the first time they saw it. Now, we will transition to a more challenging integer multiplication problem, -4×-2. The expression -4×-2 is more challenging than -2×3 because there are two negative integers and no positive integers. With -2×3, students like Gertrude can use the commutative property and the "groups of" definition of multiplication because there can be "3 groups of -2." With the expression -4×-2, using the "groups of" definition of multiplication does not work in the same way because the concept of negative groups does not make sense.

Next, we share two vignettes of how Miguel and Warren tackled this challenge. Miguel and Warren each solved -4×-2 individually with a variety of tools available, including linking cubes and empty number lines, prior to formal instruction about negative integers. Their excerpts here highlight their initial thoughts about multiplying two negative integers.

MIGUEL'S STRATEGY

The following excerpt highlights Miguel's response to -4×-2. We share Miguel's solution as an example for exploring how his thinking is logical, even if incorrect.

> *Teacher: OK. Negative 8 is your answer?*
>
> *Miguel: Yeah.*
>
> *Teacher: OK.*
>
> *Teacher 2: Why? Can you explain why you got negative 8?*
>
> *Miguel: Well, because, um, it wouldn't really make as much sense for a negative multiplied by a negative to equal a positive. It's like, um, I'm not sure how to . . . It just wouldn't make as much sense. Because if a positive multiplied by a positive would equal a positive, then I would assume that it would be the same for a negative. And it would be a negative times a negative would equal a negative.*

Miguel solved -4×-2 by making an analogy; he compared -4×-2 to $4 \times 2 = 8$. He reasoned that if a positive number times a positive number is a positive number, then it makes sense a negative number multiplied by a negative number would result in a negative number. This is how Miguel determined that $-4 \times -2 = -8$.

PAUSE AND PONDER

What do you see as powerful in Miguel's work?

_____
_____
_____
_____
_____
_____
_____
_____
_____
_____

THE POWER IN MIGUEL'S REASONING

Miguel's reasoning was entirely logical, even if not mathematically correct, because it drew on sense making. He reasoned through the comparison of signs (a positive number multiplied by a positive number compared to a negative number multiplied by a negative number). This reasoning in pursuit of sense making drew on structure, a powerful way of thinking mathematically. Indeed, a positive number times a positive number is positive, and a negative number plus a negative number is negative. It seems logical to extend that reasoning.

Making analogies with new content—here, with integer operations—is a useful strategy and works well with addition and subtraction. Young students often productively solve an addition problem such as $-3 + -2 = -5$ by comparing it to $3 + 2 = 5$ (Bofferding, 2010). Using analogies like this also applies to subtraction; for example, when the minuend has a larger magnitude than the subtrahend, or when $a > b$ in $-a - -b$, analogies from positive numbers can be extended to negative integers. Students reason that $-8 - -2 = -6$ can be compared to $8 - 2 = 6$ (Bofferding & Wessman-Enzinger, 2017). Because addition and subtraction analogies can be productive for thinking and learning about negative integers, multiplication analogies may also be helpful (such as comparing -4×-2 to $4 \times -2 = -8$ or $-4 \times 2 = -8$). Using analogical reasoning is powerful, as it supports sense making and structure, even if it resulted in an incorrect solution during Miguel's first time solving -4×-2. Analogical reasoning is a powerful mathematical tool (English, 1998, 2004), and Miguel drew on it.

For encouraging analogical reasoning, consider asking students, "Does this problem look similar to something we have already done?" or "Is there something we have already done that could help us solve this problem?"

WARREN'S STRATEGY

Warren correctly solved -4×-2 in his first attempt exploring negative integers. Multiplication with two negatives, such as -4×-2, is challenging, and students may think the way Miguel did. Although Warren's solution was correct, his strategy, which we call *countering*, is difficult to make sense of (see figure 2.2, page 32).

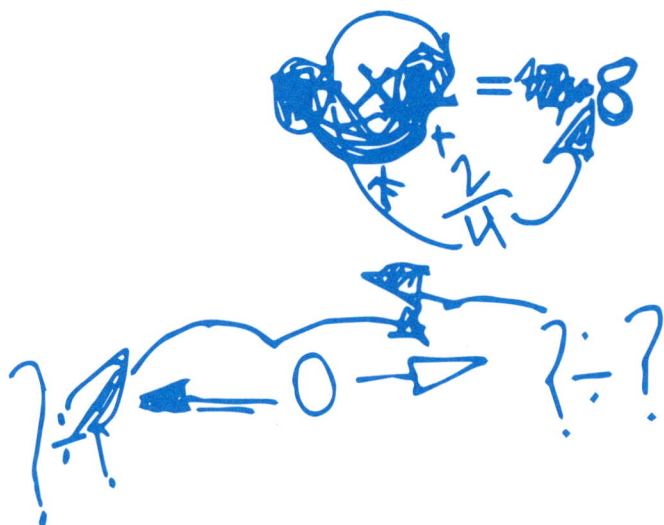

Figure 2.2: Warren's drawing of his strategy for −4 × −2 = 8.

Warren: Um, so I subtract this negative symbol. Scratch out this one and this one . . . counters this one too. So, it takes both these out, and then it's just 4 times 2. I just thought that since they're both negative numbers and that one's whole, it's basically kinda like dividing except you're multiplying. And you kinda just . . . I just thought that you counter them. It's kind of like dividing because instead of making this, um, negative 8, you just make it a normal 8, which means that the number, if these numbers were whole numbers—well, no, since they're negative numbers . . . If, like, 'cause they're not equal to zero, they're this way [left of zero], and the whole numbers are this way [right of zero]. It's kind of like since these numbers were divided, these ones we're dividing, they would get smaller or go the opposite way—like, if these ones, they would go this way.

Warren reasoned that he could ignore the negative symbols in −4 × −2 (for example, "subtract this negative symbol," "scratch this one and this one") and just think of 4 × 2. Although some procedures or algorithms that are taught advocate for "ignoring symbols," Warren did not have prior instructional experiences with negative integers, particularly procedures for them. Therefore, it seems Warren invented this strategy using conceptual understanding that he could divide out −1—for example, −4 × −2 = (−1 × 4) × (−1 × 2) = (−1 × −1) × (4 × 2). Therefore, Warren can think of just −1 × −1. Warren's explanation, "I thought you can just counter them," supports the idea of thinking about multiplying by −1, which has the effect of changing direction (see the arrows pointing left and right that he drew and his explanation about movement). Multiplying by –1 moves the product one direction, and then multiplying by –1 moves the product the other direction, which returns it to the original direction (a "counter").

THE POWER IN WARREN'S REASONING

Warren determined a correct solution for this notoriously challenging type of negative integer multiplication problem. Warren's correct determination that $-4 \times -2 = 8$ *surprised* us because he had no formal instruction in negative integers. Although Warren mentioned "scratching out" the negative symbols, this was not because of a previously learned algorithm.

Rather, Warren, evoking creative power through the pursuit of sense making, reasoned about moving left of zero and then moving right of zero—inventing ideas of countermovements (see figure 2.2, page 32). His scratching-off sign was a way to account for moving in one direction and then another. His strategy focused on the concept that multiplying by the first negative number changes the direction of movement ("go the opposite way"), and then multiplying by the second negative number changes that direction again.

BETTER TOGETHER WITH BOTH CORRECT AND INCORRECT SOLUTIONS

Whether correct or incorrect, Miguel and Warren showed powerful mathematics. Miguel used analogical reasoning, and Warren conceptualized movement on a number line; both are powerful mathematical tools.

Miguel created an analogy, and generated an incorrect solution of –8 for -4×-2. Many students use analogical reasoning with negative integers (Bishop et al., 2014; Bofferding, 2012), and such analogous thinking is productive with negative integer addition and subtraction, such as in comparing $-2 + -3 = -5$ to $2 + 3 = 5$ or $-8 - -2 = -6$ to $8 - 2 = 6$. Warren used creative reasoning for this challenging type of problem. Although he obtained the correct solution and invented a unique strategy, he was not able to fully articulate his reasoning.

Miguel's and Warren's strategies together could be beneficial. Miguel's use of analogy may enhance Warren's ideas of movement. For example, combining ideas of movement with analogy may point to comparing the movements of 4×2, 4×-2, and -4×2 to -4×-2. Representing Warren's ideas of countermovements could certainly be strengthened with his number line representation and clarity in explanation; but these initial ideas present a strong foundation for continued exploration. Miguel's strategy, although incorrect, could enhance Warren's reasoning by pushing him to articulate multiplicative movement using various number sentences.

When two of your students present ideas like Miguel's and Warren's, and you think that other students could benefit from seeing them, consider having these students share their thinking, focusing on the strategy and not the solution. Then ask the class, "Could we apply anything [student A] did to help [student B]? Could we apply anything [student B] did to help [student A]?"

PAUSE AND PONDER

Where do you see beautiful and powerful mathematical mistakes?

_____
_____
_____
_____
_____
_____
_____
_____
_____
_____
_____

Concluding Remarks

In this chapter, we highlighted the beauty and power in mathematical mistakes. To do this, we purposefully selected an element of mathematics that is not often referenced as beautiful—operations with negative integers. While tessellations, fractals, symmetry, and Fibonacci numbers in nature are highly visual aspects of mathematics that are typically dubbed as the representatives for mathematical beauty, we want to help you imagine and extend your conceptions of beauty and power in mathematics.

Because you are reading this mathematical book, perhaps you have already considered mathematics as beautiful and powerful. But have you thought about mathematical mistakes as beautiful and powerful? Have you considered your students' mathematical mistakes as beautiful and powerful? Asset-based views of mistakes recognize the beauty and the power in the students' mathematics, even if this mathematics does not initially lead to a correct solution. Seeing the beauty and power in mathematical mistakes may help your students feel more comfortable to explore and enjoy mathematics. By comparing different ideas and strategies, you highlight students' beautiful and powerful mathematics, even when the solutions are mathematically incorrect. The students here had freedom in their thinking because we did not tell them how to think but only asked them how they reasoned. This provided space for the students to invent beautiful and powerful mathematics that may not always be correct—and that's OK!

Reflection Questions

Use the following questions for reflecting on the ideas in chapter 2.

1. Recall the three words you wrote down prior to reading this chapter that you associate with the phrase *mathematical mistakes* (page 21). In what ways do your words align with our words? In what ways do your words capture different ideas?

2. What is a beautiful mathematical mistake you have made or seen?

3. What is a powerful mathematical mistake you have made or seen?

4. In what ways do you recognize beautiful and powerful mathematical mistakes in your own classroom?

Chapter 2 Application Guide

In this chapter, we argued that mathematical mistakes are beautiful and powerful aspects of mathematics. Use the following application guide to connect the chapter's themes to your classroom.

Chapter Theme	Connection and Application to Your Practice
Mathematical mistakes are beautiful.	Support students in generating new ideas or inventing mathematical strategies. This means students will make mistakes. When that happens, reflect on the order and limitations of mathematics. Embrace any surprises that happen. Instead of telling students whether they are right or wrong, ask them why what they are finding is surprising.
Mathematical mistakes are powerful.	Support students in sense making about mathematics. As they make sense of mathematics for themselves rather than being told how to think, they will make mathematical mistakes. When that happens, use the mistakes to deeply reflect on the mathematics.

Celebrating Mathematical Mistakes © 2025 Solution Tree Press • SolutionTree.com
Visit **go.SolutionTree.com/mathematics** to download this free reproducible.

CHAPTER 3

Factual, Procedural, and Conceptual Mistakes

BIG IDEA

Types of mistakes include factual, procedural, and conceptual. Being able to identify which type your student made, considering where they are on their learning trajectory and what the lesson goal is, helps you determine how to address the mistake.

In chapter 2 (page 21), we discussed how mathematical mistakes are beautiful and powerful. But are all mathematical mistakes beautiful and powerful? Are some mathematical mistakes more beautiful or powerful than others? Before reading this chapter, take a moment to reflect on or journal about the following questions: What types of mathematical mistakes do you think are beautiful and powerful? And, if you think some are more beautiful or powerful than others, share why.

A REFLECTION FROM NATASHA

As a mathematics education professor, I spent many semesters preparing future K–12 teachers by facilitating their mathematics content courses.

Unfortunately, due to the limited number of times I got to interact with students, I often had to decide on the fly what mistakes were worthy of being inspected in class. As much as I wanted to unpack all mistakes presented, seeing my students two times a week for fifteen weeks (where several of those class periods were taken up by assessments or presentations) did not allow for enough time to do so. How did I make sure my students felt their work, right and wrong, was valued?

By learning how to classify my students' mistakes as factual, procedural, or conceptual, I was able to maximize my time in class by presenting on mistakes that aligned with our learning goal, and thus advance my students' ideas about the concept. I began a system that made me feel I was helping all students—I corrected factual mistakes with them individually (through brief conversations in class or during office hours) or on their written work (always with an invitation to visit during office hours if additional conversation was needed). I inspected procedural and conceptual mistakes in class, time permitting, if they were ubiquitous or aligned with the learning goal. If time did not permit, I posted online about the mistakes in our course's learning management system and asked students to comment on them either virtually or at a later class session. This system worked for me, given the cadence of my course and the expectation that students participate in class as well as virtually. As you read this chapter and learn about the differences between factual, procedural, and conceptual mistakes, I hope you think about what structure might work for you as you plan to inspect these mistakes in your class.

★ ★ ★

Types of Mistakes

As there are different types of mathematical knowledge (for example, procedural and conceptual knowledge), there are also different types of mistakes. Conceptual mistakes, or mistakes that draw on both procedural and conceptual knowledge, are the most beautiful and powerful mistakes, as they provide the most opportunity for growth in mathematical knowledge. For example, consider the mistakes in table 3.1 (page 41). Not memorizing a fact and forgetting a procedure are just not as powerful as mistakes that occur when describing why something does or does not work in mathematics. The former mistakes often are not due to a fundamental misunderstanding, whereas the latter mistakes usually are and represent opportunities to move toward knowledge attainment.

Table 3.1: Definitions of Factual, Procedural, and Conceptual Knowledge and Mistakes

Mathematical Knowledge	Definition	Description of Mistake	Example of Mistake
Factual Knowledge	"Factual knowledge refers to having ready in memory the answers to a relatively small set of problems of addition, subtraction, multiplication, and division" (Willingham, 2009, p. 16).	Making an error about something that one *actually does have* factual knowledge about	An example of a *factual mistake* is when a student states that 2 + 3 = 6 because they confounded their fact for 2 × 3 = 6 with 2 + 3 = 5.
Procedural Knowledge	"A procedure is a sequence of steps by which a frequently encountered problem may be solved" (Willingham, 2009, p. 17).	Overlooking a step in a procedure or conducting a step in a procedure in a way that results in an incorrect solution.	An example of a *procedural mistake* is when a student forgets to regroup when using a standard algorithm for adding numbers 29 and 36, stating the solution is 55 instead of 65.
Conceptual Knowledge	"Conceptual knowledge refers to an understanding of meaning; knowing that multiplying two negative numbers yields a positive result is not the same thing as understanding why it is true" (Willingham, 2009, p. 17).	Making an error or obtaining the incorrect solution because a concept has not yet been fully developed or understood.	An example of a conceptual mistake is when a student interprets the equals sign as an operation rather than a relational symbol with 2 + 3 = _____ + 1. A student may think that 5 is the solution in the blank because the equals sign means "compute."

celebrating mathematical mistakes

Let's look at identifying mistakes in practice. In this scenario, fifth-grade students receive a set of data and are asked to share one number that captures the typical value from the data set. The data the teacher writes on the board are values representing the amounts of money different clubs earned during a local bake sale ($160, $215, $215, $110, $180, $190, and $20), so students must find the typical amount of money a club earned. Students in this classroom have had instruction on how to find the mean, median, and mode from a procedural perspective, and they work on the question in their table groups for fifteen minutes before the following discussion happens. As you read, put yourself in the position of the teacher and consider what might be your next steps.

Teacher: OK, let's start with Taryn's group. What did you decide represents a typical amount of money earned by the different clubs?

Taryn: My group said that we think $185 should be used since it is the middle number—it's between $180 and $190.

Teacher, writing "$185" on the board and "middle number between $180 and $190" next to the value: Thanks, Taryn and group. OK, let's hear from Audra. What did your group say?

Audra: We did the mean like we talked about the other day. The mean is $181, so that's our guess.

Teacher, writing "$181" on the board and "mean" next to the value: Thanks! Let's hear one more thought. Jayden's group, what did you say?

Jayden: My group talked about the middle number like Taryn's group did, and the value that Audra got, but we said the middle number is better. We said the middle number is $180.

Teacher, writing "$180" on the board and "middle value" next to the value: OK, so what did your group say about using the mean?

Jayden: Yeah, the mean. But that one club only had, like, $20, which is, like, way less than the other clubs. So we didn't think that the mean was the best way to look at a regular value.

Teacher: Typical value? I see what you're saying. So, you think the middle number, or median, is better here?

Jayden: Yeah.

PAUSE AND PONDER

At this point in the vignette, what would you do next as the teacher? Why?

From a statistical perspective, we know that this data set is skewed by the twenty-dollar value, as means are skewed by values that are very large or small. Therefore, the median is the better way to think about a typical value. Jayden's group is correct. However, we see two other mistakes. Taryn shared that her group found the median, but they had an incorrect procedure for finding the median (splitting the difference between two values rather than finding the true middle value). Audra's group found the mean, but they calculated it incorrectly (procedural mistake). Upon examination of their work, the teacher saw that although they did the procedure for the mean correctly, they had an error in their repeated subtraction when they did long division to find the mean.

PAUSE AND PONDER

Which of these mistakes provides for the most powerful step forward in learning?

In this chapter, we use a range of examples—from kindergarten to high school—for illustrating factual, procedural, and conceptual mistakes. Additionally, although this chapter focuses on setting the foundation for each of the mistake types, we note when you can use each type and how you might move learning forward in each scenario. More ideas on this topic appear in part 2 (page 79).

FACTUAL MISTAKES

Researcher Daniel T. Willingham (2009) describes factual knowledge as "having ready in memory the answers to a relatively small set of problems of addition, subtraction, multiplication, and division" (p. 16). We extend this description by defining a factual mistake as making an error about something that one *does have* factual knowledge about. An example of a factual mistake is when a student accidentally inverts digits and writes 7 + 5 = 21 when they know 7 + 5 = 12 (factual mistake). Some kindergartners may do this because they have conceptual misunderstandings about place value, but after this mistake was presented to them, they laugh and say, "Oops, I wrote it backward." This indicates that they truly understand 7 + 5 = 12, and they simply made a factual mistake in writing their solution.

We all make mistakes, and sometimes these factual mistakes happen. For kindergartners and even older students, a factual mistake might look like inverting digits. It might also be accidentally writing down the wrong number or a slight arithmetic error within a procedure. (It is important to note that the assumption here is the student knows their addition facts but may still be working to develop understanding of place value.) As the expert teacher of this student, your job is to determine what the student does or does not know to ascertain what type of mistake this represents. The general rule for a factual mistake is to determine whether, at one point, the student indicated understanding of the topic. This may look like reviewing the mistake with the student and asking them to explain their reasoning, or reviewing similar examples from classwork or assessments. If you find an indication of understanding, you can assume they have at least some knowledge about the idea.

As factual mistakes come up in class during group work or discussions, you, as the facilitator, must decide whether to share and inspect the mistakes. Researchers Alyson E. Lischka, Natasha Gerstenschlager, D. Christopher Stephens, Jeremy F. Strayer, and Angela T. Barlow (2018) provide some general rules for determining whether to inspect a mistake. We highlight two of them here.

1. **Does the mistake align with the mathematical learning goal?** Making sure that the mistake aligns to the learning goal

Factual, Procedural, and Conceptual Mistakes

of the lesson is the first step toward finding an inspection-worthy mistake. What can you learn about a student's mathematical thinking from a factual mistake in relation to the learning goal? A factual mistake is likely one of two things: (1) it may just be a simple oversight, and the student actually does understand the mathematics, (2) or it may represent something deeper that is actually related to procedural and conceptual understanding, which we will dig into later in this chapter.

2. **Is the mistake pervasive?** If more than one student or a large portion of the class is presently making the mistake in your classroom, or you know from experience that will happen, then it is a pervasive mistake worth inspecting. Pervasive mistakes happen so often because they are logical to the student (even if not correct). These types of mistakes are also rooted in procedural and conceptual understanding. We will talk more about community or collective mistakes in chapter 4 (page 66).

When a factual mistake emerges during class, you should take time to unpack it if it hinders movement toward the learning goal or if it is pervasive.

In our experience in K–12 education, factual mistakes meeting one or both rules rarely occur. Typically, we can sort out factual mistakes by working individually with students to help them recognize their errors. The following vignettes show what this may look like. We first share an example with fractions from fifth grade and then share an example with geometry from high school. If a mistake does hinder movement toward the learning goal or is pervasive, we encourage you to follow the pedagogical moves presented in the Conceptual Mistakes section (page 53).

FIFTH-GRADE EXAMPLE: FRACTIONS AND FACTUAL MISTAKES

Fifth-grade students are working on multiplication of fractions, specifically a whole number times a proper fraction. The task presented is as follows.

> *Millie is working in her garden and pulling out vegetables that are ready for harvest. Three days this week, she has been able to harvest ¾ of a pound of carrots each day. How many total pounds of carrots has Millie harvested this week?*

celebrating mathematical mistakes

Students are working on this problem in pairs or trios. The teacher, Mr. F, has asked them to model their work on paper or with manipulatives of their choice. As students work, Mr. F circulates and observes the work. Mr. F notices that Keran and José have the following written on their paper: "three groups of three-quarters is $3 \times ¾ = ⁶⁄₄$." The following interaction happens.

> *Mr. F: Keran and José, can you tell me a little about your work here? (Points to their equation.)*
>
> *Keran: José and I thought about this problem as multiplication. We know that she has three-quarters of a pound of carrots on 3 days, so that would be 3 groups of three-quarters. So that's easy.*
>
> *José: Yeah, that's just 3 times three-quarters.*
>
> *Mr. F: I see exactly what you did there. Could you please draw me a picture or use manipulatives to show your answer of ⁶⁄₄? I want to see it in pictures as well as symbols.*
>
> *(Keran and José begin to draw ¾ using a set model and represent 3 groups of ¾. Mr. F checks on other groups during this time, then makes his way back to Keran and José.) All right, I see a picture here. Tell me about it.*
>
> *José: Well, we know ¾ would look like a bar into fourths but with 3 of those fourths shaded, meaning we have 3 of the 4.*
>
> *Keran: Yeah, so then we drew that picture 3 times since it's 3 groups of ¾.*
>
> *José: Yeah, which is 3, 6, 9 groups of fourths, so ⁹⁄₄.*
>
> *Mr. F: I see. So does that match your equation? (Points to the equation with ⁶⁄₄ as the answer.)*
>
> *José: No, but we can fix that!*

What likely happened when Keran and José were working was that they confounded multiplication of 3×3 with addition of $3 + 3$ because, when prompted, Keran and José produced a correct picture representing the three groups of ¾ with a final answer of ⁹⁄₄. They recognized that 3×3 is 9 but simply made a mistake with their equation, which was illuminated in their representation. Notice that Mr. F did not quickly call out the error; that would represent recognizing and correcting the mistake. Instead, he saw an opportunity for the pair to verify their work through representations and recognize and correct their own mistake, thus reducing the likelihood that they will make this mistake again. José and Keran will hopefully remember that by verifying or checking their work,

Factual, Procedural, and Conceptual Mistakes

they can fix any factual mistakes. Additionally, notice that Mr. F did not share this mistake with the class, because it would not help move the class toward the learning goal.

HIGH SCHOOL EXAMPLE: GEOMETRY AND FACTUAL MISTAKES

Geometry students have learned about the Pythagorean theorem and are exploring how to apply that equation to finding unknown side lengths of right triangles. Mrs. C provides the students with the following task toward the end of the unit that includes the Pythagorean theorem.

> The quad at Geometer College is a quadrilateral-shaped, grassy recreational area on the college campus. It is surrounded by (that is, in the center of an open portion of) the mathematics building. Although students often walk along the quad's perimeter, especially when it is raining, there is a worn path along the diagonal that measures 15 meters. You know that one of the sides of the quad measures 9 meters. What is the measure of the other side length?

Students sit at tables of four and work with their shoulder partners on the task. Given that students already know the origin of and conceptual foundation for the Pythagorean theorem, Mrs. C wants this to purely be an application problem, where students have opportunities to practice substituting values into the equation and solving for the unknown. She encourages students to draw a picture to support their work.

Mrs. C is rotating through the classroom when she notices that Asa and Robyn have the following written on their paper: "$x^2 + 9^2 = 15^2 \rightarrow x^2 + 18 = 30 \rightarrow x^2 = 12$." The students do not have a supporting diagram for their work. The following interaction occurs.

> Mrs. C: Asa and Robyn, can you tell me a little about your work here? (Points to their equation.)
>
> Robyn: Asa and I know from what we've been talking about that 15 meters has to be the longest side. So that's why 15 goes there. And then 9 has to be one of the other side lengths, so it goes on the other side of the equals sign.
>
> Asa: Yeah, then you just plug those values in and figure it out.
>
> Mrs. C: Let's say "substitute" instead of "plug in." Remember, we want to be precise in our mathematical language. And I see exactly what you did. So, after you substituted the values, you said that 9 squared is 18 and 15 squared is 30. Remember when we drew pictures last week to prove the Pythagorean theorem? Can you draw a picture for this specific scenario, please?

(Asa and Robyn draw a right triangle and label the side lengths appropriately. Mrs. C checks on the other pair at the same table while they draw.) All right, I see a picture here. Tell me about it.

Asa: This is the quad. The diagonal is 15 meters. One side length is 9 meters. The variable x is the other side's length, and we have to solve for that.

Mrs. C: When we drew these pictures last week to prove this theorem, what else did we show?

Robyn: Oh! We drew the squares off each side of the triangle, and then we showed how the area of the two squares off the sides is the same when added together as the area of the square off the hypotenuse.

Mrs. C: Right, and how do we find the area of the square?

Robyn and Asa: Side times side!

Mrs. C: But then your area for these squares wouldn't be 18 and 30, would it? Can you verify this for me? I'm going to check on some other students, and I'll return.

(Returns after approximately five minutes.) What did you discover?

Asa: We made a mistake. We did 9 times 2 and 15 times 2 instead of 9 times 9 and 15 times 15, like the area of a square.

Mrs. C: I love that mistake and how you used the picture to help you figure it out. Keep working. Great job!

We see that Asa and Robyn confounded finding the square of a number with multiplying that number by 2, which is indicated by Asa and Robyn's production of a correct picture representing the Pythagorean theorem and correctly finding the values for this scenario. They recognized the areas of squares with side length 9 and side length 15 are 81 and 225, respectively, but simply made a mistake with their equation, which was shown in their representation. Notice that Mrs. C did not quickly call out the error and correct it by saying something like, "That's not how we find the square of 9 and square of 15." Instead, she saw an opportunity for the pair to verify their work through a representation and connect this representation back to the conceptual foundation they laid the week before. This gave the pair a chance to recognize and correct their own mistake, thus reducing the likelihood that they will make this mistake again. Asa and Robyn will hopefully remember that verifying or checking their work gives them an opportunity to fix any factual mistakes.

Factual, Procedural, and Conceptual Mistakes

> When your students present factual mistakes, consider if they are pervasive. If so, explore them as a whole class. If not, consider assisting the students individually so as not to distract the whole class from the learning goal.

From these two vignettes, we cannot ascertain whether the mistakes were pervasive. If they were, the teachers could follow an approach like the one you will see in the Conceptual Mistakes section. Assuming they were not pervasive, Mr. F's and Mrs. C's pedagogical decision to have their students work on their mistake within their pair did not diminish their work but actually strengthened their learning, as it pushed them to connect their equation to a mathematical representation of their choosing.

PROCEDURAL MISTAKES

Willingham (2009) defines a procedure as "a sequence of steps by which a frequently encountered problem may be solved" (p. 17). We use this definition of a procedure to think about what a *procedural mistake* might look like, such as overlooking a step in a procedure or conducting a step in a way that results in an incorrect solution. This type of mistake happens after students have been taught the steps for the procedure and either are practicing that procedure or will do so later, after they have learned it. An example of a procedural mistake is when a student forgets to regroup when using a standard algorithm for adding numbers 29 and 36, stating the solution is 55 instead of 65.

Returning to Lischka and colleagues' (2018) rules for determining whether to inspect a mistake, you have two potential directions to take when a procedural mistake arises.

1. If the type of procedural mistake is pervasive or aligns to the learning goal for the day, it is worthy of taking the time to inspect it during class.

2. If the mistake does not meet those two rules, it is safe to proceed to work with the individual student or group of students to help them correct their error.

As the previous section described how to help students adjust a factual mistake in their small group (rather than sharing it with the whole group), the following vignette describes how to address a procedural mistake that is pervasive (present in several groups' work).

THIRD-GRADE EXAMPLE: ADDITION AND PROCEDURAL MISTAKES

Students are learning how to add two three-digit whole numbers. The task presented is as follows.

> *Twins Kayah and Kelah have combined their money to buy a new TV for their room. Kayah has $256 and Kelah has $194. How much money do they have altogether to buy a new TV?*

Students are working on this problem in pairs or trios. The teacher, Ms. Z, has asked them to solve the problem using a picture or manipulative and then verify that answer with the standard algorithm. As students work, Ms. Z circulates and observes the work. She notices that several groups of students have correctly modeled the solution but are not getting the same answer when they perform the procedure. Although some students have written the sum correctly as $450 for the procedure, she sees incorrect answers, such as $31,410 and $540. Some students can't complete the procedure at all. The following interaction happens.

> *Ms. Z, to the whole class: I want everyone to take twenty seconds to finish what they are writing, drawing, or saying. (Pauses.) Let's come together as a whole group. I noticed that many of us have one answer in our picture and another answer in our procedure. Let's try to figure out what happened. Layla and Azriel, can you please show us the picture that you created for this problem?*
>
> *(Layla and Azriel walk to the document camera and show their picture.)*
>
> *Layla: We drew 2 hundreds, 5 tens, and 6 ones for Kayah's money and 1 hundred, 9 tens, and 4 ones for Kelah's money. Then we added it all up.*
>
> *Ms. Z: How did you add it, Azriel?*
>
> *Azriel: We said that we have 3 hundreds, 14 tens, and 10 ones.*
>
> *Ms. Z: Nice! Is there another way we could represent that amount? Emma, your group had basically the same picture, but your final sum was a little different. What did you say your final sum was?*
>
> *Emma: We said that 10 ones would have to be a new ten. So, that's actually 15 tens. But 15 tens could be made into those flat pieces, the hundreds. So really, what you have is 4 hundreds, 5 tens, and 0 ones.*
>
> *Ms. Z: Thank you, Azriel and Layla. Thank you, Emma's group. Many of you had similar pictures and sums to these two groups'. However, some of us didn't get the same answer when we checked it with the procedure. I saw one group write the procedure as 256 + 194 = 31,410, and I see exactly how they got that answer! I want us all to talk about*

this way to solve it with the procedure. Does it match our picture? What might be going on? Take a few minutes in your group to talk about this way to solve it with the procedure, if you agree with this approach, and how it connects to our picture.

PAUSE AND PONDER

If you were Ms. Z, how would you have approached the situation? What would you have done next?

_____
_____
_____
_____
_____
_____
_____
_____
_____
_____
_____

Let's pause and unpack a few things in this vignette. Ms. Z noticed that, although most students could represent this sum in pictures, many failed to use the standard American algorithm correctly. As they are still early in the learning trajectory for adding two three-digit numbers, Ms. Z is not surprised nor concerned. Also, since these procedural errors are pervasive and they align with the learning goal, she has asked the students to inspect a mistake by sharing it with the whole class. She made this decision on the fly during instruction, as she sees the opportunity to connect the mistake to deeper learning. Additionally, notice that Ms. Z decided to present the pervasive mistake herself as a general one, rather than calling on a particular group to share their procedural mistake. This was purposeful. Ms. Z has been working toward students' recognition of mistakes as mistakes—shifting them from a deficit-based view to an asset-based view of mistakes. Not all her students are yet comfortable or willing to share when they have incorrect solutions. Rather than potentially embarrassing students by drawing attention to them and what they view as a mistake, Ms. Z decided to share the work herself as an opportunity for learning.

> If you are still building your students' ability to confidently share mistakes, consider first sharing mistakes generally. You do not want to embarrass or discourage students; so, one strategy is not to attribute mathematical mistakes to a particular student or group at first.

After asking her students to discuss the mistake presented, Ms. Z takes time to facilitate a discussion on how the mistake makes sense. A few groups of students clearly articulate where all the digits in the answer originated. Let's return to Ms. Z's class.

Ms. Z: Kara, can your group explain what happened in this mistake?

Kara: Yeah, it looks like what happened was they added the numbers, so 6 plus 4 is 10, and then 9 and 5 is 14, and 2 and 1 is 3. Then they just wrote 3 next to 14 and 14 next to 10. They didn't actually regroup.

Ms. Z: Class, when Kara says "9 and 5 is 14" and "2 and 1 is 3," where do you see those sums in the pictures we saw earlier? Take twenty seconds to talk about that in your groups. (Students talk in groups.) Zion, I heard your group talking, and I want you to share what you said with the whole class.

Zion: We said that it's not really 14 or 3. It's really like saying 90 and 50 is 140, and 200 plus 100 is 300. But the person who did that mistake was really close—they just wrote down those numbers but didn't think about that 31,410 is 31,000!

Ms. Z: OK. So what I'm hearing is that this person with the mistake added correctly. We can verify that with our picture. But they wrote down those sums for each hundred, ten, and one, but didn't yet think about the place value or regrouping.

Zion: Yeah, I think so.

Ms. Z: Let's go back to this idea of regrouping and connect what we've learned about regrouping in the procedure to the picture we saw earlier.

Ms. Z continues by having a minilesson on how to represent regrouping in the standard American algorithm for adding, connecting each regrouping to the picture presented earlier. After this minilesson, Ms. Z asks the groups to talk to their partners and revoice how they would do the standard procedure for this problem (not yet putting pencils to paper). After students revoice, she calls on a few groups to share out and then asks them to retry the original problem.

Factual, Procedural, and Conceptual Mistakes

> For a pervasive mistake, pause the whole group and ask students to make sense of the mistake, similar to what Ms. Z did. Additionally, ask students to verify what was done well or how the student could have made the mistake. This approach encourages the asset-based view of mistakes, as it celebrates the sensibleness of the mistake that was made.

From this vignette, we hope it is clear how procedural mistakes, although not conceptual, still lead to rich discussions that develop conceptual understanding. In our early years of teaching, we often made the error of simply correcting students who made the kind of mistake presented in the vignette. However, after years of shifting our thinking to asset-based views of mistakes, we recognize this formerly quickly dismissed mistake is actually a rich foundation on which we can discuss the standard American algorithm for adding digits and connect that to representations using pictures and manipulatives.

CONCEPTUAL MISTAKES

Conceptual knowledge in mathematics matters because it helps students understand why they do what they do (NCTM, 2014). Willingham (2009) defines conceptual knowledge as "an understanding of meaning; knowing that multiplying two negative numbers yields a positive result is not the same thing as understanding why it is true" (p. 17). We can see how this is different from procedural knowledge, which focuses on mechanics rather than meaning. Therefore, we define a *conceptual mistake* as making an error or obtaining the incorrect solution because a concept has not yet been fully developed or understood. We see this in an example where a student is given $2 + 3 = \underline{} + 1$ and thinks what goes in the blank is 5. This indicates that the student views the equals sign as an operating symbol (similar to the plus and minus signs) instead of a relational symbol. This isn't a procedural mistake, because the student has not failed to understand or properly execute a procedure. Instead, the student has not yet developed the conceptual underpinnings needed to understand relational symbols.

How do you identify a conceptual mistake versus a procedural mistake, and when do you decide to share and inspect conceptual mistakes in the classroom? To identify these mistakes, you must give students opportunities to present conceptual mistakes. Problems or tasks that aim for correct answers without providing opportunities for open-ended strategies or solutions and time to justify one's reasoning are unlikely to elicit conceptual mistakes. Tasks should encourage students to share their strategy and why the strategy they chose works—this can include written or verbal justification as well as pictures and representations. These tasks should also move away from procedural solutions only and strive to be open-ended or open-middle tasks. Open-ended tasks have more than one solution; open-middle tasks have more than one way to solve them (even if they have one solution).

Next, we provide a vignette of what this may look like in your classroom.

KINDERGARTEN EXAMPLE: ADDITION AND CONCEPTUAL MISTAKES

Students are learning to add single-digit numbers and complete number sentences with two addends on either side of the equals sign. The task presented is as follows.

> *Cre has 5 apples, and she needs to put them into 2 different baskets. She has found a way to do this and writes a number sentence to represent her grouping: 1 + 4. Write a number sentence that represents another way Cre could group her apples that is equivalent to what she has found.*

Students are working on this problem in pairs. The teacher, Mrs. G, has written "1 + 4 = ___ + ___" on the board and has given the pairs blocks for representing the apples to assist in their number sentence writing. As Mrs. G circulates through the room, she notices that a few pairs have some correct solutions, but at least three pairs have an incorrect solution (such as "1 + 4 = 5 + 1"). The following interaction happens.

> *Mrs. G: I want everyone to pause in their work. I am going to write a few solutions that I see in your groups on the board.* (Writes one correct solution, 1 + 4 = 2 + 3, and one incorrect solution, 1 + 4 = 5 + 1, on the board.) *What I want you to do in each of your groups is to use your blocks to show each of these sentences and then decide: Are they equivalent?*
>
> (Students begin modeling the sentences. After students have had a chance to model each of them, Mrs. G calls the whole class back together.) *I want to hear what you and your partner discovered! Nia and Bobby, what did you two discover?*
>
> *Nia: We think that they are the same* (Mrs. G politely inserts the appropriate vocabulary of "equivalent"), *equivalent, because both of them have 5 apples.*
>
> *Bobby: Yeah. Cre still has 5 apples to split up. So, both of those on the board still have 5 in each basket.*

PAUSE AND PONDER

How would you handle this interaction? What would be your next teacher move?

_____
_____
_____
_____
_____
_____
_____
_____
_____
_____

Let's take a step back from Mrs. G's classroom. Your initial reaction—and ours when we first began our teacher journeys—might have been to correct Nia and Bobby. We can imagine a teacher saying, "No, one of the sentences is not correct, and one is." However, if we stop and think about the conceptual reasoning behind Nia and Bobby's statement, we begin to see that their understanding of the equals sign is operational instead of relational and that, since Mrs. G saw similar misunderstandings among other groups, this mistake is pervasive.

When a pervasive conceptual mistake appears, having students share their reasoning by either presenting their work or dissecting other work allows the teacher to get a peek into their justification. This approach can prepare you to make the next instructional move that will rectify the conceptual error while celebrating the mistake.

When a conceptual mistake manifests in the classroom, teachers should rejoice! Factual mistakes and instruction that focuses on developing facts have long prevailed in mathematics classrooms, and students know this. They are prepared to make factual mistakes. However, conceptual mistakes represent something deeper and more important in terms of knowledge acquisition. Many students have not been given opportunities that will elicit such mistakes. So, when those mistakes appear, pat yourself on the back! It means that your instruction is giving students the time and opportunity to develop a deeper level of knowledge than simple fact or procedure acquisition.

Let's return to Mrs. G's classroom and see how she handles this situation in a way that celebrates the mistake without diminishing the work and reasoning that went into Nia and Bobby's understanding.

> *Mrs. G: Nia and Bobby, thank you for sharing that. I see exactly what you mean. In both of these sentences, on either side of the equals sign, we have* at least (adding emphasis with her voice) *5 apples. You all know that I love hearing all different kinds of ideas, so I want to hear from someone who has a different perspective. Malea and Zade, can you share what you two discovered?*
>
> *Malea: We don't think they are the same* (Mrs. G again politely says "equivalent"), *equivalent, because the problem says that Cre only has 5 apples. One of the sentences has 6 apples when you use the blocks. Five can't be 6. So, they can't be the same.*
>
> *Mrs. G: Zade, do you agree? What do you want to add?*
>
> *Zade: I think in the second sentence, that person plussed* (Mrs. G politely says "added"), *added, 1 and 4 to get 5 but then wrote the 1 to make it look like the left side. But that's not what the equals sign means. The equals sign means they have to be the same on both sides, and 1 plus 4 is not 5 plus 1.*
>
> *Mrs. G: OK. So you think the equals sign is like the scale we have in our classroom?* (Pulls out the scale and begins putting 1 and then 4 blocks on one side and 5 and 1 blocks on the other.) *To make this balance, or to be equal, then, what is on one side must be equivalent to or balanced with what is on the other side. Since 1 plus 4* (points to the scale) *is not balanced with 5 plus 1* (points to the scale), *then these can't be equivalent statements. Does that jibe with what you're saying, Zade?*
>
> *Zade: Yes! Just like when we used the scales!*
>
> *Mrs. G: I want to see what everyone thinks about that. Take out a sticky note and write your name on it. Then, on your way out the door to lunch, I want you to put the sticky note with your name on it next to the number sentence that you think is correct. If you think both are correct, then you need two sticky notes, both with your name on them. We will revisit this after lunch.*

We want to point out some important things about Mrs. G's instruction. Mrs. G never once identified the number sentences on the board as incorrect or correct. Additionally, as students were sharing, she did the

Factual, Procedural, and Conceptual Mistakes

same with student thinking; she did not label student thinking as correct or incorrect. Mrs. G told Nia and Bobby, "Thank you for sharing that. I see exactly what you mean." She thanked them for their mathematical contribution and acknowledged the sense in their reasoning, even though it was wrong. This may seem like a small thing, to thank a student, but it encourages the student to share their reasoning, regardless of correctness. In this way, Mrs. G demonstrated to her students that *all* thinking is valued.

Also, several times in the vignette, students used mathematical language that was not exactly appropriate. Mrs. G has made it a practice to gently nudge students toward the correct vocabulary; however, she does not do this by negatively calling them out. This practice is something she has done with these students since day one, so they do not find her intrusion of their speech disruptive. We encourage you to consider adding such a tactic *after* you have worked on community norms with your students so they learn that this is expected and not evaluative (Parrish, 2011). Additionally, by using the scale (a manipulative the class is familiar with), Mrs. G emphasized the conceptual meaning of the relational symbol, equals, again without labeling thinking as correct or incorrect. This encouraged students to think of ways to use tools and representations to verify their reasoning.

When conceptual mistakes occur, think of ways to encourage students to use tools, representations, or manipulatives to deepen the thinking and solidify the concept.

While her students are at lunch, Mrs. G reviews the sticky notes and sees the conversation has moved many students to the correct solution, but she has two pairs that still have incorrect thinking. She decides to do a minilesson with these four students during intervention that has more direct instruction involved. She will remind these students, with the use of the scale, that the equals sign relates two statements by saying they are equivalent, or balanced, rather than being operational (telling them to *do* something). Then, she will ask students to use the scale to show some other statements that are equivalent to 1 + 4.

As you saw with Mrs. G, we encourage you to address any conceptual mistakes that are present in the classroom as soon as possible. If the conceptual mistake aligns with the learning goal for the day, then the mistake can be addressed during instruction. If the mistake does not align with the learning goal, you do not necessarily need to address it that day, but you should take time to meet with that individual student to work with them. This could occur during intervention or small-group instruction time. Regardless of when or how the conceptual mistake is addressed, we want you to make every effort to address it, as conceptual mistakes have greater repercussions on subsequent learning than factual mistakes (which have historically taken the stage in classroom error analysis or correction). Since factual mistakes often represent slips or misremembrances, these mistakes are usually remedied by the teacher, the class, or the students themselves. Given the hierarchical nature of mathematical learning, if students do not have opportunities to correct conceptual mistakes, they might not understand subsequent topics (Kshetree, 2018).

Concluding Remarks

In this chapter, we presented three different types of mathematical mistakes: (1) factual, (2) procedural, and (3) conceptual. Being able to identify which types of mistakes your students are making will give you insight into your next teaching moves. Perhaps you will notice a conceptual mistake and decide to design a task the next day in class that incorporates that idea. Or, maybe you will think nimbly on the spot within a whole-class discussion and use a conceptual mistake as a way to explore mathematical ideas more deeply. Think about ways this classification of mistakes can support decisions you make in your classroom.

Reflection Questions

Use the following questions for reflecting on the ideas in chapter 3.

1. Did this chapter shift your thinking about the types of mathematical mistakes you consider and discuss in class? How and why?

2. Select a mathematical content area that you teach (for example, mean or operations with fractions). Using that content area, create examples of a factual mistake, a procedural mistake, and a conceptual mistake.

3. Consider the following example where the students were tasked with finding the perimeter of the shaded shape in figure 3.1 (page 59). How would you classify these mistakes? Do you think one of these mistakes is more discussion worthy than the other? Why or why not?

Factual, Procedural, and Conceptual Mistakes

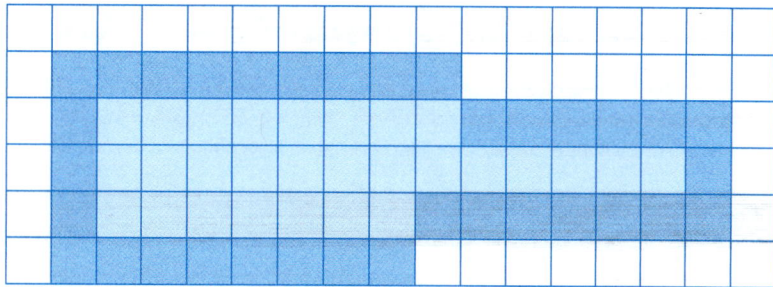

Figure 3.1: Shape for determining perimeter using the darker shaded part.

Anders: For the perimeter, I am not sure if I should count the edges or the boxes. Maybe I should count the darker boxes to find the perimeter because they do go around the shape.

Bernadette: I added up all of the sides (9 + 6 + 3 + 7 + 8 + 5 = 38) because you add up all of the sides for perimeter.

4. What is a conceptual mistake that you made as a learner or that you have seen your students make that stands out to you? Why is this conceptual mistake memorable? How did it help you or the students learn?

Chapter 3 Application Guide

In this chapter, we shared about three types of mathematical mistakes. Recognizing these mistakes helps you select which ones you want to investigate with an individual student or with a whole class. Use the following application guide to connect the chapter's themes to your classroom.

Chapter Theme	Connection and Application to Your Practice
Factual mistakes	A factual mistake may be a simple slipup (the student does understand the concept) or may indicate deeper conceptual misunderstandings. Often, these mistakes can be corrected one-on-one with a student or in a small group and do not need a whole-group focus.
Procedural mistakes	A procedural mistake involves overlooking a step in a procedure or conducting a step in a way that results in an incorrect solution after students have been taught the procedure's steps and are still practicing. Determine whether the procedural mistake is pervasive or aligns with the learning goal before determining how to address it.
Conceptual mistakes	A conceptual mistake involves making an error or obtaining an incorrect solution because a concept has not been fully developed or understood yet. A conceptual mistake is a rich area to inspect and use for student learning. You should always consider how to encourage inspection of and reflection on these types of mistakes.

CHAPTER 4

Mistakes by Mathematicians

BIG IDEA

We are all mathematicians. All mathematical mistakes that help people learn are worthy of celebration. This chapter is celebratory of different mathematical mistakes. This explicit celebration is a tool for supporting the development of your and your students' identities as mathematicians.

> Before reading this chapter, take a moment to reflect on or journal about the following questions: How do you celebrate mathematical mistakes? Do you consider yourself a mathematician? What makes a mathematician a mathematician?

A REFLECTION FROM NICOLE

The other day, I picked up my daughter from a community center art class. As I greeted an art teacher who was checking children in and out with parents, the teacher asked me, "Oh, are you the mathematician here to pick up your daughter?" My initial internal reaction was, "No! I'm not a mathematician." My second internal reaction was, "Wait, why do I feel this way?" To the art teacher, I hesitated and replied, "Well, it depends on your philosophical position on who and what a mathematician is." Embarrassingly, I actually started describing the work of a research mathematician and how, though I love mathematics and teaching mathematics, I consider myself more of a mathematics enthusiast than a mathematician. At this, the art teacher merely laughed and said, "Yes, you are the mathematician."

As I held my daughter's hand and left the community center, I thought about how I think of my child as a mathematician, and I think of my students as young mathematicians. I see the readers of this book as mathematicians. Yet I do not think the same thing of myself. I see the irony in that. Even as someone who does mathematics and teaches and writes about mathematics every day, I hesitate to call myself a mathematician.

What does this have to do with mistakes? Mathematicians make mistakes, revise their mistakes, and then make more mistakes. Similarly, those of us who teach mathematics will make mistakes, revise our mistakes, and then make more mistakes. If we cannot see ourselves as mathematicians, then we run the risk of seeing the work we do as unworthy of being held to the same standards as that of mathematicians.

PAUSE AND PONDER

Do you see yourself as a mathematician? Do you see your students as young mathematicians? Why or why not? What types of mathematician activities do you and your students do?

_____
_____
_____
_____
_____
_____
_____
_____
_____
_____

In this chapter, let's celebrate the mistakes of *all* mathematicians. Celebrating mistakes is the zenith of all asset-based discussions and reflections on mathematical mistakes. Imagine a continuum from young mathematicians' to professional mathematicians' mistakes. We will celebrate awesome mistakes along this continuum in this chapter.

> When you normalize mistakes as the work of mathematicians, you can help support your students' identities as mathematicians. Valuing your students' mistakes with either words of affirmation or a mistake at the center of a lesson is a way to celebrate mistakes and treat your students as mathematicians.

A Continuum of Favorite Mistakes

In this section, we explore different mistakes across a continuum of mathematicians as a way to celebrate mistakes. Celebrating mistakes is a deliberate action to counter deficient narratives of wrongness in mathematics. We first celebrate mistakes in society to provide a historical view of mistakes. We then give insights into mistakes by celebrating favorite mistakes from research mathematicians. And finally, we consider favorite mistakes from our students (who are also mathematicians).

PAUSE AND PONDER

What is your favorite mathematical mistake?

_____
_____
_____
_____
_____
_____
_____
_____
_____

MISTAKES AND SOCIETY: THE CASE OF ZERO AND NEGATIVE NUMBERS

Can you imagine doing mathematics without zero or negative numbers? Imagine the challenges of dealing with numbers if you did not have symbols or a way to conceptualize zero or negative numbers. The first recorded zero we have is from the Sumerian culture in Mesopotamia, about five thousand years ago or 3 BC (Kaplan, 2000; Nieder, 2016). The Mayans also independently invented zero about AD 4. Historically speaking, discovering or inventing zero was a big deal for mathematics and society because it helped with developing positional numeral systems and considering other numbers, such as negative numbers.

Even with the advancements of conceptualizing zero, the development of the number system as we know it today took extensive time. In fact, mathematicians struggled with the concept of negative numbers for centuries. For example, in the 3rd century, Diophantus, who is often called the "father of algebra," described equations like $x + 20 = 4$ as absurd (Bishop, Lamb, Philipp, Schappelle, & Whitacre, 2011). Think about that; the father of algebra thought negative numbers were absurd! Great minds of the past struggled with concepts like zero and negative numbers for centuries, and students today are expected to make sense of these topics within weeks. Recognizing that humans made mistakes about numbers, particularly zero and negative numbers, for centuries is hopefully liberating to think about. Certainly, it is empowering to give grace to yourself and your students who wrestle with the complexities of mathematics.

MISTAKES AND COMMUNITY: THE CASE OF 0.9999 . . .

In our experience as mathematics teachers, asking a student to share a personal mathematical mistake before you develop an environment of celebrating mistakes can lead to feelings of shame or closed-off communication. The individualistic focus on mistakes is counterproductive to the goal of moving toward asset-based views of them. When experiences shift from individualistic to communal, it can be easier to talk about incorrect solutions or mistakes. For example, in our work, we find that when we give students a choice, they prefer to talk about a mistake the entire class made, rather than just their personal or individual mistakes. As a teacher, you can reflect on ways to make group mistakes transparent. You can also think about strategies that elicit group mistakes. One of those strategies is reflecting on favorite mistakes.

In one of our university courses for future teachers, we implemented a final project called "My Favorite Conceptual Mistake," which we share more about at the end of this chapter (page 73). Essentially, the project encourages students to reflect on their mathematical experiences over the semester and present on their favorite conceptual mistake. After several years of facilitating this final project, we noticed our aspiring teachers often focused on mistakes that nearly the entire class made—collective, rather than individual, mistakes.

One common group mistake centers on the question, does 0.999... equal 1? Most of our students, no matter the grade level (from middle school students to university-level teachers), state that 0.999... does not equal 1 when asked if 0.999... = 1 is true, which we mentioned in chapter 2 (page 22). Aspiring teachers have shared not understanding 0.999... equals 1 as their favorite mistake every year that we have implemented this project. For example, in one presentation, an aspiring teacher shared about how the course started with students voting on whether they thought 0.9999... equaled 1. That particular year, twenty-three of the twenty-six students initially stated no; two initially said yes because "you can round it to 1"; and one student shared they did not know. Our student presented this mistake as one the entire class made. They shared how thinking about $\frac{1}{3} = 0.3333...$ and $\frac{1}{3} + \frac{1}{3} + \frac{1}{3}$ shifted their thinking. They concluded the presentation with three proofs of why 0.999... equals 1. This has happened in other ways and with other tasks as well.

Highlighting a pervasive mistake or the mistake of a community creates a safe space for students where they can find favorite mistakes to celebrate. Use these mistakes to establish the expectation that mistakes are valuable, expected, and respected.

One of our students, Emily, talked about the idea of collective mistakes in a focus group interview with us. She shared that she had chosen a collective mistake from an algebra problem because she had made a lot of mistakes in class and was still trying to understand some of them. Additionally, she had chosen one that she felt comfortable sharing in front of the class. She had used our collaborative class notes and found an algebra problem where most of the class had made a mistake. About the mistake, she said, "I think I was one of the few people at my table that thought that the volume would not be conserved, so I felt confident that I would be able to describe the mistake" (E., personal communication, April 2018). Emily helped us think about how focusing on a collective mistake, rather than an individual one, normalizes mistakes and empowers students as they change their thinking about new mathematical ideas.

In another focus group interview about mistakes, Rudy also shared about normalizing mistakes in a group. "I make a lot of mistakes. You make mistakes all the time in mathematics. If you can learn from it, which you almost always can, it is beneficial for what you made the mistake on" (R., personal communication, April 2018). Michael, agreeing with Rudy, shared, "It's important for the teacher to not put you down for your mistake and to help you figure out what you can do better" (M., personal communication, April 2018).

Norma, another student, shared something similar: "We make mistakes all the time. . . . We make them almost every day and then we fix them." She continued, "Now that I have had Nicole's class, I am a lot more open to mistakes. I don't even like to call them mistakes. I like to call them learning experiences. It's not like we are making a mistake; we get to learn from it. So, what I learned from Nicole's class is that it's not bad to make a mistake" (N., personal communication, April 2018). In this sense, Rudy, Michael, and Norma normalized mistakes as the acts of a community by describing them as things that everyone will make and learn from. Furthermore, Emily pointed out how it's easier to make mistakes as a group than individually.

After you identify the pervasive mistakes that you encounter in your mathematics lessons and you celebrate the mistakes of the community by discussing them in class, you can deepen mathematical understanding by revising the community mistakes as a whole group. This correction can be student led or teacher led, depending on the needs of your classroom.

MISTAKES FROM RESEARCH MATHEMATICIANS: FERMAT'S LAST THEOREM

The Pythagorean theorem, $x^2 + y^2 = z^2$, essentially states that with a right triangle, the square of the hypotenuse is equal to the sum of the squares of the triangle's other two sides. This theorem is ubiquitous in high school mathematics and has hundreds of ways to prove and think about it. And an infinite number of solutions (x, y, z) make $x^2 + y^2 = z^2$ true.

But would this type of theorem hold for other integer degrees beyond x^2, y^2, and z^2? Is $x^3 + y^3 = z^3$ true? Is $x^4 + y^4 = z^4$ true?

Fermat's Last Theorem states there are no natural numbers greater than 2 (3, 4, 5, . . .) that would make $x^n + y^n = z^n$ true. Likely, Pierre de Fermat wrote this around 1637 (Kilani, 2023; Miner, 2013; Muzundu, 2024; Singh, 2002). And this theorem remained unproven for centuries, until Andrew Wiles, a research mathematician, proved it in 1994 (Kilani, 2023; Miner, 2013; Singh, 2002). It was an incredible feat and a now infamous story within the mathematics community. It may be hard to imagine this, but a mathematical mistake (and likely many) is embedded in this story.

Andrew Wiles's (1995) incredible feat proving Fermat's Last Theorem is sometimes described with language like "his first successful proof." This inherently means he had other unsuccessful proofs. In fact, Wiles worked in secret on the problem for six years. He first presented a proof on Fermat's Last Theorem in 1993. However, a mistake was found, and he had to go back and continue working on this proof for a year (Miner, 2013). Wiles almost gave up, thinking he could not find the error. On the morning of September 19, 1994, as he reflected on why his approach could not work, he had a revelation about how he could prove this theorem, and he did so (Singh, 2002). Proving Fermat's Last Theorem is an exemplar of how even the absolute best research mathematicians struggle and make mistakes. Growing into and doing the work of a mathematician entails embracing struggle and mistakes.

As we prepared for this book, we sent out a survey to professional mathematicians, asking them to tell us their favorite mathematical mistake. We also set up individual interviews with career mathematicians. One professorial mathematician wrote in her survey, "My favorite mathematical mistakes are when I'm with a group of people and we realize our mistake and then find other ways we could have thought about the problem" (J. Long, personal communication, August 2023). That reminded us of the ways our students often reference collective mistakes. When we asked this mathematician why she favors this kind of mistake, she responded:

> This is my favorite because I'm with a group of people who [are] revising or revisiting a problem or scenario we thought we understood, and then thought of it in a completely new way. Usually, this new way we wouldn't have encountered had we not made a mistake. It leads to more interesting discussion and understanding that we would not have had otherwise. (J. Long, personal communication, August 2023)

Similar to our earlier discussion, this research mathematician reflected on the social or community aspect of mistakes and the joy that can bring.

The research mathematician shared that it took her fifteen years to get her mathematical dissertation research published. In those years of revising her work and submitting it to different journals, she encountered doubt. And she said she still experiences unsureness:

> I will say it's something I continue to struggle with. I didn't publish my dissertation in a peer review journal until fifteen years after I graduated. I have two to three papers that are in the revising stage right now. So, they actually haven't appeared . . . and I graduated fifteen years ago. . . . I wondered if this isn't even a paper, do I even deserve my degree? . . . In one feedback, [a reviewer] said this should not be published anywhere ever. . . . I just really internalized "I'm not good enough to do this." . . . For me, as a teacher, those feelings that I had myself are the same things my students go through. When my students say, "I'm not good at [this]. I'm not a math person. I can't do this," that's the same version of [me saying], "I shouldn't have graduated, I don't know how to write papers, and I don't have what it takes," that I still struggle with. (J. Long, personal communication, August 2023)

When the research mathematician shared this, we thought immediately about our students over the years (elementary students to university-level students) who have experienced doubt, had low confidence, and struggled with their mathematical identities. In mathematics, there has been a focus on confidence and identities (Boaler, 2016) because these can be obstacles to student learning. Doubting oneself and feeling insecure about mistakes seem to be universal experiences for all mathematicians, from the young mathematician to the research mathematician. We wondered if any of our students knew that research mathematicians, even ones with PhDs and careers, have felt this way. Of course, we did not want our students to feel this way, but we wondered if speaking about it more and normalizing that mathematicians also experience self-doubt and make mistakes would cause our students to be more open to share about their mistakes.

PAUSE AND PONDER

What are ways you can support your students in having confidence as they make mistakes?

_____
_____
_____
_____
_____
_____
_____
_____

How to Elicit and Celebrate Mistakes

By sharing favorite mistakes from multiple perspectives, we aim to paint a positive picture of mathematical mistakes. Most importantly, we hope these examples normalize mistakes for you and your students. Further, these celebrations of mistakes support the view that all learners and doers of mathematics are mathematicians.

Before we can celebrate mistakes, we need to gather mistakes from your students. Here is a list of suggestions for gathering students' mistakes.

- Use a rich mathematical task.
- Provide space and time for thinking deeply about the task.
- Support different ways of thinking and a variety of strategies.

Using a rich or interesting mathematical task is one way to elicit mistakes. Although there are various ways of thinking about richness, we envision *rich tasks* as having three attributes.

1. Rich tasks provide students multiple strategies for solving the problem.
2. Rich tasks allow for solutions that are unobvious or surprising (tying to our idea that the beauty in mathematics includes the element of surprise).
3. Rich tasks move beyond procedures and instead develop conceptual reasoning, what researchers Margaret Schwan Smith and Mary Kay Stein (1998) call "doing mathematics."

Although many resources for mathematical tasks exist—from physical curriculum to digital resources on social media—not all resources are made the same. In fact, some resources on social media are replete with poor mathematical tasks (see Hertel & Wessman-Enzinger, 2017). As such, we hope that by considering our three features of rich tasks, you can begin to filter through the tasks you have access to so you can distinguish which tasks are truly rich. This may involve solving the tasks yourself and also thinking about how your students may solve them, given their prior knowledge and experiences. We encourage you to review Smith and Stein's (1998) journal article on cognitive demand for mathematical tasks as a starting point for being able to identify rich ones.

Mathematical tasks that can be solved in more than one way (sometimes called *open-middle* tasks) can be a good start. Also, mathematical tasks where the answer is not obvious, or the answer is surprising, can be interesting. For example, the task we discussed earlier, "Does

0.999. . . = 1?" is a rich task because most students intuitively think 1 cannot be equal to 0.999. . . and are surprised when they find out it is.

Students cannot make mistakes or experience mistakes as a group or community if they do not have time and space to work and discuss collaboratively (Parrish, 2011). Create space and time for students to make mistakes and discuss them. This can mean adjusting and repositioning what your mathematics class looks like. It requires slowing down. Although you may be doing fewer problems as a result, you will be going deeper with a few good ones.

Once you have a good task and have slowed down to allow for deep thinking, you can support students in creating their own ways for solving the problem. As students share their ways of thinking with each other, new mistakes and understandings will be supported.

Interestingly, both our students and research mathematicians have described making mistakes collectively. This focus on collective mistake making, rather than individualistic mistake making, reveals something important to consider for your work as a teacher. You must give your students opportunities to work together on mathematics that allow for collective mistakes to occur. Additionally, there must be time for discourse about strategies instead of just solutions. When you pivot from a focus on only solutions to strategy discussions, you move from concentrating on an individual's mistake to learning as a community together.

Embedded within the goal of celebrating mistakes is the imperative need for recognition that we are all mathematicians, even if we are not professional mathematicians. Are you a human? Do you do mathematics? Then you are a "mathematics person." As a society, it seems normalized to hope all children can perceive themselves as artists (Boone, 2007; Mendelowitz, 1963). Even as adults, we ought to embrace our inner artist, as that has ripple effects on our well-being and self-care (Vaartio-Rajalin, Santamäki-Fischer, Jokisalo, & Fagerström, 2021). In a way, we can all be artists. It is the same with mathematics.

To fully accomplish the aim of celebrating all mistakes in this chapter, educators also need to recognize our role in doing mathematics. While we defined mathematical mistakes as beautiful and powerful in chapter 2 (page 22) and shared three types of mistakes in chapter 3 (page 40), we can also have mistakes around our mathematical identities (for example, "I am not a mathematics person" or "I have to be fast at doing problems to be good at mathematics"). We deeply hope that you not only can see mistakes as worthy of celebrating, but also can embrace your and your students' identities as mathematicians. It is certainly something we are also working on.

An Act of Celebrating

What is an act of celebrating mistakes that you could enact in your classroom? Likely, celebrating mistakes is more than one thing (see part 2, page 79) and even a philosophical undertaking. Celebrating mistakes is cheering for your students' invented strategy and seeing it as beautiful and powerful mathematics, even when it does not result in the correct answer. Celebrating mistakes is normalizing them, so much so that they become an integral element of teaching. Celebrating mistakes is at the heart of supporting your students' identity shifts by seeing them as mathematicians doing the work of mathematics. But perhaps there is one act you could do that would be the ultimate celebration of mistakes. We confronted this challenge ourselves and grappled with it together.

We thought to truly center mistakes and celebrate mistakes, we needed to have a final project that celebrated mistakes. This is how we created the "My Favorite Conceptual Mistake" project (Wessman-Enzinger & Gerstenschlager, 2023). We thought this final project would do three things.

1. We thought it would center mistakes and allow us to reference them over and over.

2. By using an asset-based word like *favorite* as an adjective with *mistake*, we hoped it would help shift mathematical views.

3. As we discussed in chapter 3, some mistakes help people learn more than other mistakes do, so we decided the project should have students share conceptual mistakes rather than factual or procedural ones.

When we implemented the "My Favorite Conceptual Mistake" project, we thought students would share procedural or conceptual mistakes. Although students did, some students shared about "mistakes" regarding how they viewed themselves as learners of mathematics and what mathematics is. Figure 4.1 (page 74) shows a drawing of how a student initially thought that mathematics is all procedures and focuses only on being correct quickly. This is stressful, like standing on the edge of a plank on a pirate ship. Then he learned mathematics is more than that, and he is capable of doing mathematics.

Source: © 2023 by Evan Carlson. Used with permission.
Figure 4.1: A student's drawing depicting how he used to see himself doing mathematics when the focus was on correctness alone.

Concluding Remarks

Doing mathematics deeply means developing perseverance and taking on challenging problems. As such, mistakes are unavoidable. In fact, making mistakes is an inherent attribute of the work of a mathematician. In this chapter, we connected how making mistakes and revising them is something that mathematicians do. We highlighted the work of a prominent mathematician and a research mathematician, demonstrating the pivotal role of mistakes in mathematics. In fact, one of the mathematicians we interviewed shared, "What a mistake even is, is something that is evolving for me" (J. Long, personal communication, August 2023). We hope that to be true for you in this chapter. Is your definition of who a mathematician is evolving? Is your definition of what a mistake is evolving?

Reflection Questions

Use the following questions for reflecting on the ideas in chapter 4.

1. In what ways are you already celebrating mistakes by the mathematicians in your classroom?

2. How might you celebrate the mistakes of the mathematicians in your classroom?

3. What mathematical mistakes related to your own identity as a mathematician have you made?

4. How do you think you can support your students in seeing themselves as mathematicians?

Chapter 4 Application Guide

In this chapter, we shared how seeing yourself and your students as mathematicians normalizes the role of mistakes. Embracing and celebrating mistakes for learning not only helps your students learn but also supports their mathematical identities. Use the following application guide to connect the chapter's themes to your classroom.

Chapter Theme	Connection and Application to Your Practice
Celebrating mistakes	Celebrating mathematical mistakes centers the mistake as something worthy that helps learning. Centering mathematical mistakes as valuable tools is part of the work of a mathematician. Find ways to highlight mistakes from others (famous mathematicians, other teachers, you, and other students) regularly and in a celebratory fashion so students begin to normalize and value mistakes.

PART 2

Responding to Mathematical Mistakes in Action

In part 2 (chapters 5–10, pages 83–186), we provide explicit and tangible pedagogical suggestions for incorporating and supporting mathematical mistakes in your classroom in asset-based ways. These chapters are examples of task structures that use mistakes to leverage learning forward. As you engage in each chapter, envision what these might look like in your own space.

CHAPTER 5

Two Foundational Instructional Strategies for Examining Mistakes

A foundation for investigating mathematical mistakes in your classroom can begin with using "unknown student work" (Barlow, Gerstenschlager, & Harmon, 2016) that contains a mistake and carefully selecting tasks.

> Before reading this chapter, take a moment to reflect on or journal about the following question: In what ways does your current classroom allow for investigating mathematical mistakes from an asset-based perspective?

A REFLECTION FROM NATASHA

Early in my teaching career, I regularly showed examples of student work that included mistakes I wanted my students to inspect. I envisioned that this presentation would include a lively discussion of the mistakes from an asset-based perspective and would support students in making sense of the problems. At the time, these examples were from actual students.

Much to my chagrin now, I realize how this practice diminished students' confidence (and thus was why none of my students wanted to engage with the mistakes in the way that I had envisioned). Therefore, I switched to presenting the mistakes as ones I made. I would say to my students, "I tried solving this problem, and I made a mistake in my work. Can you help me figure out what I did?" Although this change in my approach enticed a few more students to participate, I still did not achieve full class participation. Additionally, I missed participation from some groups of students entirely, particularly students from marginalized communities and students struggling in class.

★ ★ ★

Research has indicated that students from nondominant groups (racial, cultural, linguistic, and so on) often become ostracized from learning opportunities, while students who have characteristics of the dominant group are provided greater access to learning opportunities simply by dominating group discussions (Kurth, Anderson, & Palincsar, 2002). This was exactly what was happening in my class! Those students who always participated continued to do so. Those who I really wanted to reach continued to escape me. Further, when students did participate, they weren't convinced that what I was presenting was a mistake. Many believed there was no way I, as the teacher, would make a mistake (given my role of power and authority as the teacher, I should have recognized this as a potential issue!).

Finally, after years of making this mistake in how I presented mistakes, I reframed the ones I shared as what my colleagues and I call *unknown student work* (Barlow et al., 2016). Rather than presenting work my current students did or something I did, I would tell my students the following when I presented work on the projector that included a mistake: "This came from one of my students last year. You might know them! I won't tell you who, but they said I could share their work because the mistake they made was *so* great and clever." The shift in participation was astounding. Students were ready to rip apart the mistake—in good ways! They found how the unknown student's work could be used to find the right answer and ways to help the student take the next step from the work shown. From that point on, unknown student work was a permanent part of my tool kit! As you read this chapter, think about how unknown student work and carefully selected tasks can make their way into your tool kit as well.

In this chapter, and for the remainder of this book, we aim to provide explicit tools for supporting mistakes in your classroom right away. We begin with laying the foundation for an asset-based perspective, using unknown student work, and selecting tasks or revising existing tasks to promote investigation-worthy mistakes.

Laying the Foundation for an Asset-Based Perspective

An important aspect of laying this foundation is understanding why an asset-based perspective is needed to move learning forward. This requires first expanding on what we mean by an asset-based perspective, which we introduced in chapter 1 (page 12).

Coming to a classroom with an *asset-based perspective* means anticipating and looking for strengths in varying facets of humanness that make students unique people (honoring cultures, languages, neurodiversity, socioeconomic statuses, immigration statuses, and so on). It means seeing your students as humans who add value to your mathematics classroom in distinctive ways. We recognize that this approach to teaching has many other names (for example, *culturally responsive teaching* and *complex instruction*), but we use *asset-based perspective* to encompass all the varying terms. And we view *asset-based pedagogy*, such as culturally responsive teaching and complex instruction, to be a way of teaching that supports asset-based perspectives on students and their strengths. Research shows that asset-based perspectives support mathematics development for all students, but particularly for those traditionally left behind by previous pedagogies (Celedón-Pattichis et al., 2018). Let's discuss why is this type of pedagogy and approach to learning is particularly important as you consider mistakes.

For a student who may already lack confidence in mathematics, treating mistakes from a deficit-based perspective diminishes that student's thinking, thereby "dehumanizing" the student and producing "suboptimal learning" (Adiredja & Louie, 2020, p. 43). Researchers Nicole Louie, Aditya Adiredja, and Naomi Jessup (2021) coined the term *deficit noticing* to describe situations where teachers attend to (or notice) mistakes of students from marginalized communities and attribute these mistakes to a lack of knowledge, or deficits in the students' backgrounds. This behavior ignores the limited opportunities students might have had to develop such knowledge. Naturally, being treated in this manner by a teacher can lead a student to feel inadequate or dehumanized. And naturally, this experience is awful and does not position the student as a powerful mathematician.

Instead, teachers should recognize that all students have logical conceptions about mathematics. That logic may not align to the system of mathematics adults use in their curriculum or classroom; but students' logic is relevant and real in its own beautiful and powerful way. As such, treating mistakes from an asset-based perspective means that teachers see all students as having logical conceptions. We consider a logical conception to be when a student is employing sense making or structure, or leveraging their prior knowledge. Although it may not always result in the "correct" mathematical solution (or what the mainstream mathematics community views as correct), we can understand the student's sense making or logic.

Asset-based perspectives mean that teachers will recognize the variety of mistakes students make by considering their varied backgrounds and identities and how those influenced their arrival at the mistakes. Treating mistakes in this way allows for all students to feel engaged in the learning because everyone's conceptions are valued, respected, and worthy of investigation. For example, a fourth-grade student from Norway may write decimals differently than North American students do. In Norway (and other countries), 23.53 is represented as 23,53, using a comma as the decimal point. This notation is not incorrect; it's just different, and collaborative class discussions can address other ways to write numbers.

Similarly, consider a second grader whose Mexican grandmother teaches her a subtraction algorithm, one that is extremely efficient but does not match the standard algorithm that is taught in U.S. schools. This algorithm is different from the one used in class, so maybe the second grader forgets a step in the class's algorithm or makes a mistake. The other algorithm or idea should not be overlooked, but rather embraced and discussed. The literature indicates the importance of highlighting the ways mathematics is used and valued in various cultures, particularly ways that are counter to the traditional white and Western positions of mathematics taught in nearly all American schools (see Battey & Leyva, 2016, for a thorough description of whiteness in mathematics education).

There are many nuanced steps in striving to become an asset-based educator. Belief systems need to be dismantled, curricula need to be re-envisioned, and systemic structures need to be destroyed. Our intent in this chapter is not to discuss all these ideas, because many brilliant educators and researchers offer wisdom in those ways (Hammond, 2015; Muhammad, 2020; Song & Id-Deen, 2023; Ukpokodu, 2011). Our intent is to provide a few pedagogical suggestions that you can use to immediately start making a shift to asset-based education. These pedagogical practices may take time to become second nature and require some general communication and discussion norms to be in place in your classroom.

Two Foundational Instructional Strategies for Examining Mistakes

You may use the following list of sample classroom norms to make this asset-based shift at any grade level.

- Listen to who is presenting their mathematical ideas, and keep your eyes on their mathematics.
- Disagree agreeably ("I disagree because . . ."; "I understand why you think that, but I am thinking about . . .").
- If you are confused, ask for clarification ("Can you repeat that in a different way?" "I was confused when you said . . ." or "Can you explain that again?").
- If you agree and would like to add to what was said, do so ("I agree because . . ."; "I would like to add on to that").
- Focus on the mathematics being presented and not on the presenter or the aesthetics of the presentation (for example, comments about a color the presenter chose for their diagram are not pertinent).

Also, Lischka and colleagues (2018) provide additional resources for establishing classroom norms for productive discussion about mistakes. Do not rush yourself! Does any of the preceding norms seem like something you want to try? Plan to learn and try ways of implementing norms that fit for you.

Early in the school year, have a conversation with your students about classroom norms, and create your own version to post as an anchor chart. By contributing to the anchor chart, students become invested in the norms and are more likely to respect and remember them.

In addition, give yourself grace and be kind to yourself. We have made many mistakes on our journey toward investigating mistakes with an asset-based perspective, and you will too. Although it may seem redundant to say, we must say it: accept the mistakes you make on this journey, and learn from them. In fact, take a moment to inspect them yourself!

Inspecting your own mistakes is beneficial in this process. For example, do you consider yourself a mathematician sharing the joy of mathematics? If you struggle with your own mathematical identity, then you might not be able to maximize the view of your students as mathematicians. Have you made a mathematical mistake and found value in it? If you have not developed your own asset-based views of mathematical mistakes, then you may not be able to elevate your own mathematical environment to its full potential. Forming an asset-based perspective toward you, your students, and your classroom facilitates creating an environment that supports mistake making.

The following sections contain two foundational instructional strategies for examining mistakes. First, we present an instructional strategy that allows *all* students to engage in investigation of mistakes without feeling shame or disenchantment because their personal mistakes do not make it to the document camera for inspection. This pedagogical strategy is called "using unknown student work" (Barlow et al., 2016). The second strategy we share involves consideration of the types of tasks you select so that the likelihood of investigation-worthy conceptual mistakes increases, but the risk to students remains minimal.

Using the Unknown Student Work Strategy

Using unknown student work is a great gateway instructional strategy in that it works well when you are just getting started with using mistakes in asset-based ways, when you are starting a new academic year, or when you have students who might be particularly shy or hesitant to share their work for inspection. This strategy allows students to feel empowered and engaged in discussion of a mistake because the conversation centers on the mistake and the learning that can come from that mistake. Yet students are still able to make sense of others' reasonings. NCTM (2014) suggests to "elicit and use evidence of student thinking" (p. 53) as a way to make instructional experiences. Using unknown student work first utilizes student thinking in a safe way so that students may be more apt to share their own thinking later. To illuminate this strategy, let's consider a pair of examples that involve fraction division and probabilities.

EXAMPLE OF UNKNOWN STUDENT WORK IN THE DIVISION OF FRACTIONS

A teacher, Mr. T, has asked his sixth-grade students to consider the problem $7\frac{1}{2} \div \frac{1}{3}$. His class has been exploring the division of fractions for a few days, and this problem is something most students find accessible.

After students have worked in pairs on their answer, but before any student presentation of work has happened, Mr. T pauses the class and says:

> Now that we've worked on this problem for a bit and many of you have solutions to the problem, I want us to take a minute to inspect the following student work. Focus on the mathematics you see and not anything like, "I don't like the marker colors they chose." This work is from my student from last year; let's call him K. T. I want to give you

Two Foundational Instructional Strategies for Examining Mistakes

twenty-five seconds of silent time just to look at K. T.'s work. Don't start discussing it yet. I will give you time in a moment. (Sets a timer for twenty-five seconds, and projects the work in figure 5.1.) Now turn to your shoulder partner and discuss this work. What do you notice? What do you wonder? Remember, we are thinking like mathematicians!

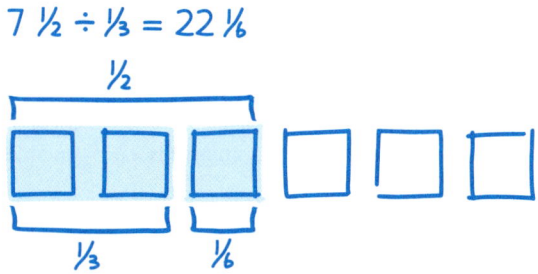

Figure 5.1: K. T.'s work.

PAUSE AND PONDER

Why might Mr. T have decided to share work from a "student from last year"? What opportunities does presenting the work as such afford in terms of discussing a student's mistake?

_____
_____
_____
_____
_____
_____
_____
_____
_____
_____

In this example, a student is solving 7 ½ ÷ ⅓. The student correctly found the whole number of one-thirds that can divide seven and one-half (22). They then drew a picture to represent the ½ found in the dividend, and they counted how many times ⅓ (the divisor) fits into that ½ (noting this with the shading). When they did that, the student misinterpreted the leftover piece of ⅙ to be part of the solution, rather than considering how many times ⅓ could then evenly divide ⅙. Our experience and others' research (Petit, Laird, Marsden, & Ebby, 2016) indicate that this is a very difficult concept for students to understand. Additionally, this concept moves instruction toward the goal of deeper understanding of fraction division, represents a pervasive mistake among fraction division, and is a fundamental misconception of interpretation of the remainder (Lischka et al., 2018). As such, this mistake is worthy of celebration and inspection in that it leads to deeper understanding of fraction division.

Here, the concept K. T. is struggling with is how to interpret the leftover $1/6$. Since $7\ 1/2 \div 1/3$ can be thought of as $(7 \div 1/3) + (1/2 \div 1/3)$, K. T. correctly divides the whole number by $1/3$ but struggles to interpret the remaining $1/6$ piece. (Note the correct answer is $22\ 1/2$ since $1/6$ is half of $1/3$.)

When presenting unknown student work, you can have the work prepared before the lesson so that it is easy to share during class time.

So why not use one of Mr. T's current students to share this work? There are two valuable reasons to use unknown student work in this situation. First, none of Mr. T's students may have come up with this mistake (which is a conceptual mistake and thus worthy of inspection). In that case, you may say, "Well, if no one has this mistake, why bring it up at all?" Because for this mistake (that is, misinterpreting the remainder), understanding what went wrong and how to correct it sets the foundation for conceptual understanding of the fraction division algorithm (what many call *keep-change-flip*), and the understanding can be extended to division of all types of numbers, thus deepening that understanding. This mistake is conceptual in nature and would be valuable for all students to inspect and understand.

Second, even if one of Mr. T's students came up with this mistake, the classroom dynamic may not be conducive to exploring that student's work. For example, maybe the student is shy or had a bad day, or maybe the classroom norms aren't quite established to allow for inspection of mistakes to be productive and focused on mathematics rather than the person or aesthetics. By seeing this mistake as that of a student unknown to them, the class can be critical of the mathematics without fear of hurting someone's feelings. Finally, since this work came from a student, they may be more likely to look for an error, rather than assuming it is correct, which is what students typically do if teachers present erroneous work as coming from themselves.

Two Foundational Instructional Strategies for Examining Mistakes

> Consider using unknown student work at the beginning of the year when students might not yet be comfortable sharing their mistakes or when a mistake does not naturally present itself in class and you know it would be valuable to inspect.

EXAMPLE OF UNKNOWN STUDENT WORK IN PROBABILITIES

Before we look at another example of an unknown student's work, let's consider the rules for a mathematical game that involves rolling two six-sided dice. In this game, pairs of students will roll the two dice and take the positive difference (for example, rolling a 2 and a 5 is a difference of 3). One student is a winner if they get the low differences: 0, 1, and 2. The other student is a winner if they get the high differences: 3, 4, and 5.

Most of our students who have played this think it is a fair game because each student gets three differences. Consider the sample space (the set of all possible outcomes) of the 36 ways of rolling two dice and the differences between the outcomes on the individual dice in figure 5.2. In this example, you can see that the probability of obtaining the low differences of 0, 1, and 2 is 24/36. The probability of obtaining the high differences is 12/36. Although there are three differences for each player to win, these differences (that is, outcomes) are not equally likely.

	1	2	3	4	5	6
1	0	1	2	3	4	5
2	1	0	1	2	3	4
3	2	1	0	1	2	3
4	3	2	1	0	1	2
5	4	3	2	1	0	1
6	5	4	3	2	1	0

Figure 5.2: Sample space for the differences when rolling two dice.

PAUSE AND PONDER

How would playing this game without first having a sample space make it challenging for students to determine whether this game is fair? How would creating a sample space be challenging for students?

_____
_____
_____
_____
_____
_____
_____
_____
_____
_____

When we use this game, we do not tell students the game is unfair (that is, the probabilities of winning are not equal). We let the students play the game and see what happens. As they play, they encounter unfairness; while sharing the results of playing, they will notice all the players with low differences tend to win. As the students start to question the game's fairness, we ask them how they can prove whether it is fair. This encourages students to create a sample space, which our students have often struggled with. Because this is memorable for us, we are prepared with unknown student work. After years of teaching this, we know half the class will eventually make a sample space with 36 possibilities, like figure 5.2, although not always in an array or grid. The other half will actually make a sample space with 21 possibilities (see figure 5.3, page 93).

For the 21 possibilities, students often justify that they do not want to "double-count" the possibilities of rolling two dice. For example, if one rolls a 2 with a pink die and then a 5 with a yellow die, the difference is 3. If one rolls a 5 with a pink die and then a 2 with a yellow die, the difference is still 3. Of course, the correct answer is that these are two distinct events. But many of our students regularly struggle with this because they confuse the rules of the game (that is, there's one way to subtract) with the sample space itself (that is, there are two separate outcomes). That being said, the piece of unknown student work in figure 5.3 is ideal for us as we anticipate this way of student thinking. Therefore, we bring it to class, ready for discussion. Making this sample space the center of the discussion confronts and centers

Difference	Combinations	Total
0	6-6, 5-5, 4-4, 3-3, 2-2, 1-1	6
1	6-5, 5-4, 4-3, 3-2, 2-1	5
2	6-4, 5-3, 4-2, 3-1	4
3	6-3, 5-2, 4-1	3
4	6-2, 5-1	2
5	6-1	1/21

Figure 5.3: Unknown student work for a sample space for the differences when rolling two dice.

the mathematical ideas. Although there will inevitably be mathematical disagreement, those who are "wrong" will have to shift their thinking and those who are "right" will need stronger justifications.

PAUSE AND PONDER

What's a piece of student work from your own classroom that is memorable to you? Why is it memorable? How can you incorporate this memorable work as unknown student work in your future teaching? (Think about how you will present it, what questions you will ask about the work, and so on.)

Selecting Tasks to Promote the Inspection-Worthy Mistakes Strategy

A second instructional strategy is carefully selecting tasks that promote the likelihood of students' making inspection-worthy mistakes in such a way that they are not discouraged from sharing. These tasks align with recommendations from NCTM (2014), which encourage teachers to "implement tasks that promote reasoning and problem solving" (p. 17). Let's start this conversation by considering the two tasks from an introductory algebra class (typically taken in eighth or ninth grade) in table 5.1.

Table 5.1: Comparing and Contrasting Tasks A and B

Task A	Task B
At the local carnival, it costs $5 to enter and $1 to play each carnival game. If Syra wants to play 10 different games with her friends, how much money should she bring?	At the local carnival, it costs $5 to enter and $1 to play each carnival game. Using the slope-intercept form, determine how much money Syra should bring if she plans to play 10 games.

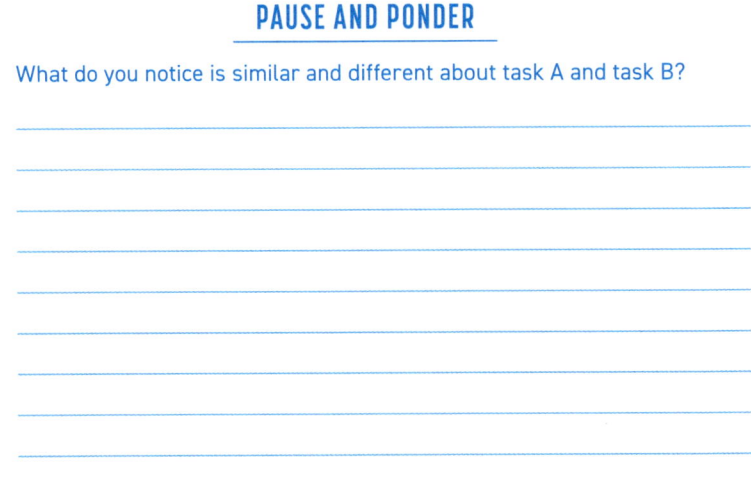

PAUSE AND PONDER

What do you notice is similar and different about task A and task B?

Two Foundational Instructional Strategies for Examining Mistakes

In these tasks, the similarity is that we know there is one answer to the problem. As such, we would identify these as closed-ended tasks. However, there is a distinct difference between task A and task B regarding *how* students are directed to solve task B using the slope-intercept form. Why is this important?

Task A is what we would call an *open-middle task* (Sheffield, Meissner, & Foong, 2004), as we anticipate that students will solve this in a variety of ways based on their knowledge (for example, direct substitution into slope-intercept form, a table, a graph on the coordinate plane, and guess-and-check). Meanwhile, task B, which we can call a *closed-middle task*, asks students to solve the problem in one specific way (use of the slope-intercept form).

Open-middle tasks provide fertile ground for conceptual mistakes. However, closed-middle tasks, by dictating the strategy for solving, typically only yield procedural mistakes. Although procedural mistakes are still worth celebrating, they are often not as rich as conceptual mistakes in terms of moving toward the learning goal. Now, let's consider the two tasks shown in table 5.2. Similarly, we want you to compare and contrast these two tasks.

Table 5.2: Comparing and Contrasting Tasks C and D

Task C	Task D
The honor society is hosting a school dance to raise money for the annual academic competition trip. It plans to charge an entrance fee of $5 per person. It will also sell drinks at the dance for $1 each. If 100 students RSVP to the dance, how many drinks will the honor society have to sell to make its goal of $1,000 for the trip?	The honor society is hosting a school dance to raise money for the annual academic competition trip. It plans to charge an entrance fee (no more than $10 per person) and sell drinks at the dance. If the honor society wants to raise $1,000 for the trip, how much should it charge for the entrance fee, and how much should it charge for the drinks? Note—you will have to make some assumptions!

PAUSE AND PONDER

What do you notice is similar and different about task C and task D?

Notice that tasks C and D, unlike task B, are open-middle tasks. Students are not encouraged to use a particular method and likely will produce various strategies to solve the problems. This variety of strategies is excellent, providing even more opportunities for inspection-worthy mistakes. When students create a variety of strategies, that means there is room for creativity and mistake making. Both creative and innovative ideas, even if not fully completed, will be great options for inspection-worthy mistakes.

Task C is an open-middle (no strategy offered) but closed-ended (only one answer) task, so students may focus on the answer alone rather than the strategies. In contrast, task D has multiple answers (especially based on the assumptions students make)! Multiple answers will create a space where students really have to focus on their peers' thinking and the strategies, rather than the solution alone.

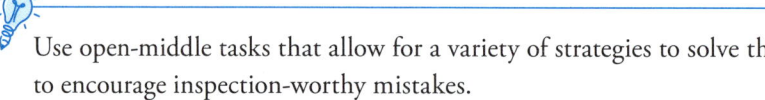

Use open-middle tasks that allow for a variety of strategies to solve them to encourage inspection-worthy mistakes.

Closed-middle and closed-ended tasks have their place, particularly on assessments to determine how students can solve certain problems and indicate their knowledge of concepts. However, during the learning process, open-middle and open-ended tasks provide opportunities for students to demonstrate their knowledge and creativity. Once you begin implementing open-middle and open-ended tasks in your classroom, how do you leverage this fertile mistake-making ground in a way that respects students' knowledge and encourages them to share their mistakes? Let's see how Mr. T does just that in the following vignette.

> *Mr. T: Let's take a look at task D. Anka, can you read task D for us? (Anka reads the task.) With your shoulder partner, what I want you to do is begin solving this problem. Start by having a conversation with your partner about how you will solve the problem. Then start your strategy.*
>
> *(Before most or all groups arrive at an answer, Mr. T reconvenes class.) I noticed that many of us are getting close to a solution. What I want is for the groups that I call on to share their solution strategy. Explain to us how you decided your approach, any assumptions you made, and any aha moments you had. Remember, what we are sharing here is* not *our final solution, and as such, this represents rough-draft thinking. We are always able to change our mind!*

Mr. T then proceeds to call on carefully selected pairs of students to share their work. He sequences their presentations so that they build in terms of conceptual thinking and sophistication, but each pair he has chosen has made a mistake. Additionally, he focuses students' attention on sharing their strategy (acknowledging that since this is an open-middle task, many solution strategies will appear) and labels their work as rough-draft thinking (Jansen, 2020). By setting up the conversation in this way, Mr. T allows the class to recognize that what is presented should not be considered the final solution, and since a variety of strategies are being shared, students can concentrate on the conceptions presented and how they can improve the work.

As you shift to use this practice in your classroom, rather than creating all new tasks, first consider all the tasks you currently use (in your curriculum or from your repertoire of materials). Organize them based on whether they allow for multiple strategies (open- or closed-middle) or multiple solutions (open- or closed-ended). If you realize that you need more open-middle and open-ended tasks, there are easy ways to convert your tasks into open-middle or open-ended ones. In the next section, we explain how you can convert tasks.

Converting or Revising Existing Tasks

To convert a closed-middle task to an open-middle task, remove any language that indicates to students how they should solve the problem. For example, if the problem says, "Using the standard multiplication algorithm, solve the following problem," remove the "using the standard multiplication algorithm" phrase. Additionally, you can add language that says "using any method you wish" or "using pictures, words, and symbols" to encourage multiple methods. Table 5.3 illustrates a kindergarten task and a high school task before and after they are revised.

Table 5.3: From Closed-Middle to Open-Middle Tasks

	Before Closed-Middle Task	After Open-Middle Task
Counting task for kindergarten	Determine how many objects are in this picture by counting by 10s.	Determine how many objects are in this picture. Share how you are thinking about it!
Probability task for high school	Consider the following scenario. 1. Flip a coin three times. You win if all the flip results match. 2. Draw a tree diagram to illustrate the sample space for flipping a coin three times. What is the probability of all three coin flip results matching?	Consider the following scenario. 1. Flip a coin three times. You win if all the flip results match. 2. What is the probability of all three coin flip results matching? Share your reasoning.

To convert a closed-ended task to an open-ended task, you may have to rewrite the problem as something vaguer. Consider what the learning goal is for the problem and use that to guide the rewriting. For example, if in the past you asked students to solve the quadratic $x^2 - 6x + 9 = 0$ because you want them to know how to find the roots of a quadratic with real roots, you could instead ask them, "We know that $x^2 - 6x + 9 = 0$ is a quadratic with real roots. Does another quadratic that has the same roots as this quadratic exist? If so, what is it? How do you know it has the same roots?" Notice how these questions will meet the learning goal for the problem. But this new problem is open-ended and potentially open-middle. These questions allow for richer discussion and offer potentially more investigation-worthy mistakes. Table 5.4 (page 99) illustrates the before and after of elementary and high school tasks, revised from closed-ended to open-ended.

Table 5.4: From Closed-Ended to Open-Ended Tasks

	Before Closed-Ended Task	After Open-Ended Task
Addition task for first grade	2 + 18 = _____	Write number sentences (for example, 10 + 10 = 20) for how you can get a sum of 20.
Geometry task for high school	Calculate the perimeter of the following triangle. 8 in, 13 in	Draw different-shaped triangles that have a perimeter of 30.

Concluding Remarks

If it is the first time (for you or your students) that mistakes are being centered in a mathematics lesson, then using unknown student work can be a safe place to begin normalizing the role of mistakes in your classroom. Pairing unknown student work with carefully selected tasks that are interesting and open-middle not only engages your students in doing mathematics deeper but also moves them toward experiencing mistakes as worthy and valuable in learning. This will open the door for students to share their individual mistakes in the future. Use the reproducible "Chapter 5: Two Foundational Instructional Strategies for Examining Mistakes" to help you implement these strategies.

Reflection Questions

Use the following questions for reflecting on the ideas in chapter 5.

1. What is a practice you already do that you could integrate unknown student work into?

2. What other norms can you think of that must be in place for investigation of mistakes to be purposeful in your classroom?

3. What kinds of mathematical tasks will create opportunities to inspect mistakes?

4. Take a moment to look through some of your current student work or to reflect on a mathematics lesson that you currently teach. What is unknown student work that you could use the next time you teach the lesson? Why have you picked this piece of student work?

Chapter 5 Application Guide

In this chapter, we explored how utilizing unknown student work as an instructional strategy and task structure is an inaugural way to start using mistakes in the mathematics classroom, and how carefully selected tasks can promote investigation-worthy mistakes. Use the following application guide to connect the chapter's themes to your classroom. Next, use the reproducible "Two Foundational Instructional Strategies for Examining Mistakes" (page 101) for unknown student work.

Chapter Theme	Connection and Application to Your Practice
Unknown student work	Unknown student work is actual work from a prior class or work that you remember from your years of experience and recreate for your students. Your students will see the unknown student work as something they can relate to, as it is similar to a peer's work. Yet your students' work is not at the center. Therefore, if students are unsure about sharing their mistakes with the whole class, this can be a great starting point.
Carefully selected tasks	Carefully selected tasks are chosen to elicit certain mistakes. Open-middle and open-ended tasks allow for more opportunities for inspection-worthy mistakes. Consider ways to alter your current tasks so that they become open-ended or open-middle.

Chapter 5: Two Foundational Instructional Strategies for Examining Mistakes

Instructions: Use a conceptual mistake or an invented strategy as unknown student work. Ask yourself the following questions.

- How could the student have gotten this solution?
- What questions would you ask the student?

Mathematical task:
Student work and solution:

CHAPTER 6

Changing Minds in Mathematics

BIG IDEA

The Changing Minds Task Structure is a useful way to support the use of both incorrect and correct solutions in the same task.

> Before reading this chapter, take a moment to reflect on or journal about the following questions: When have you witnessed your students changing their minds about a mathematical strategy or solution, and what is that like? How do you currently support students in changing their minds about a mathematical idea?

A REFLECTION FROM NICOLE

When I first started teaching high school mathematics, I lectured nearly every day and had occasional activities and projects. As I transitioned from primarily lectures to a more discourse-oriented classroom, I noticed that sometimes my students asked, "What's the answer? What's the right answer? Is it OK to change my mind?" I did not want them to be embarrassed about changing their minds. Changing one's mind, revising a strategy, and taking a different strategy pathway are all part of doing mathematics. In fact, as mathematicians work on complex problems (sometimes for years), they often change their minds. Changing one's mind is the work of a mathematician.

As I grew as a mathematics teacher who supported conceptual understanding and mathematical discourse, I desired to honor changing one's mind in my classroom. So, I now often ask my students, "Has anyone changed their mind? Why or why not?" These questions support students in vocalizing their shifts in thinking and makes explicit the concept that it's OK to change their minds. Therefore, Natasha and I developed the Changing Minds Task Structure, which is also a general pedagogical tool and is discussed in this chapter.

★ ★ ★

In this chapter, we unpack the first layer of changing minds in mathematics, which is changing minds as a creative act, illustrating creativity from both professional and young mathematicians. Then, we explore the second layer of changing minds, which is the normalization of changing minds as a broad pedagogical tool. We conclude the chapter with the last layer of changing minds, which is specific actions that you can integrate into the classroom.

Changing Minds

Often as students work on mathematical tasks, they present a series of ideas, strategies, and solutions that may or may not be correct. As they make their mathematical journey through a rich task, these opportunities to change their mind about the way they approach and solve the task are important to encourage. Teachers must highlight such opportunities because the act of changing one's mind instills the sense that being mistaken is a natural part of reaching a solution. The Changing Minds Task Structure supports the use of both incorrect and correct solutions in the same task. Although the changing minds task is an explicit tool teachers can use in the classroom for supporting mistakes, we also see changing minds as a creative act and a general pedagogical tool. Figure 6.1 (page XX) shows all three of the Changing Minds Task Structure's layers: (1) creative act, (2) pedagogical tool, and (3) task.

Changing Minds in Mathematics

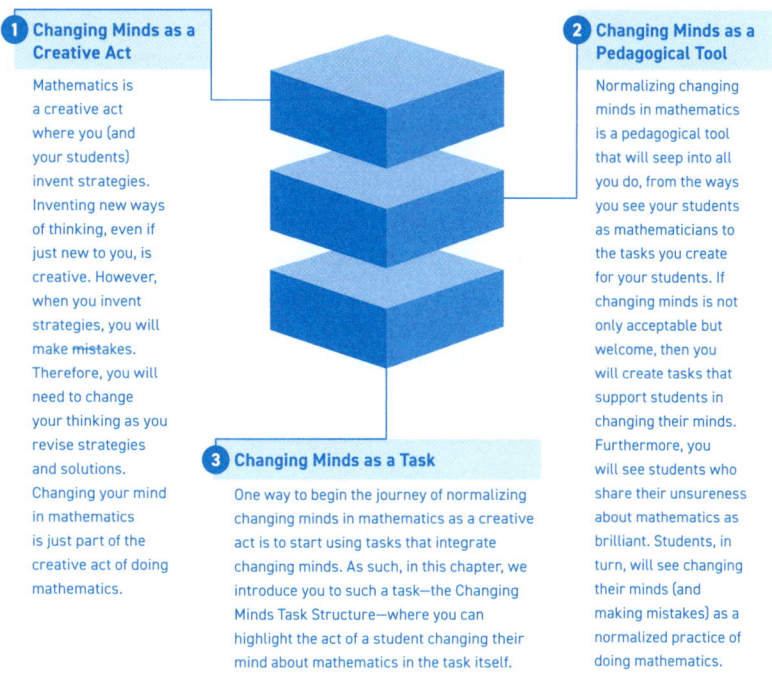

1 Changing Minds as a Creative Act

Mathematics is a creative act where you (and your students) invent strategies. Inventing new ways of thinking, even if just new to you, is creative. However, when you invent strategies, you will make mistakes. Therefore, you will need to change your thinking as you revise strategies and solutions. Changing your mind in mathematics is just part of the creative act of doing mathematics.

2 Changing Minds as a Pedagogical Tool

Normalizing changing minds in mathematics is a pedagogical tool that will seep into all you do, from the ways you see your students as mathematicians to the tasks you create for your students. If changing minds is not only acceptable but welcome, then you will create tasks that support students in changing their minds. Furthermore, you will see students who share their unsureness about mathematics as brilliant. Students, in turn, will see changing their minds (and making mistakes) as a normalized practice of doing mathematics.

3 Changing Minds as a Task

One way to begin the journey of normalizing changing minds in mathematics as a creative act is to start using tasks that integrate changing minds. As such, in this chapter, we introduce you to such a task—the Changing Minds Task Structure—where you can highlight the act of a student changing their mind about mathematics in the task itself.

Figure 6.1: Layers of changing minds.

As you reflect on what changing minds in mathematics is, as shown in figure 6.1, our hope is you will find the layer that is most helpful and meaningful for you at this moment in your teaching. Maybe thinking about changing minds as a creative act will inspire you to re-envision other areas of your mathematics pedagogy. Or perhaps you will try implementing changing minds within a task structure in your classroom and see how your students engage in thinking and learning about mathematics with it.

Changing minds is an inherent attribute of doing mathematics deeply. Furthermore, supporting opportunities where students can change their minds with mistakes fosters a safer space for them to be creative and innovative in mathematics. In the following sections, we present the three layers of changing minds, which offer flexibility for you to incorporate them into your teaching practice.

CHANGING MINDS AS A CREATIVE ACT

Let's look at Jace, a fifth grader creating his own strategies for $-5 - 4 = \_\_\_$ and changing his mind about strategies and solutions. His work is shown in figure 6.2 (page 106).

celebrating mathematical mistakes

$-5 - 4 = -1$
$-4 - -5 = 9$

Figure 6.2: Jace's solutions to $-5 - 4 =$ _____.

PAUSE AND PONDER

Why is supporting changing minds important? In what ways do you see Jace acting creatively in making mistakes and changing his mind?

_____
_____
_____
_____
_____
_____
_____
_____
_____
_____

As you read the following vignette, look for comparisons that Jace makes as he thinks about $-5 - 4 =$ _____ (for example, comparing $-5 - 4 =$ _____ to $-5 - -4 = -1$). Although Jace does not obtain the correct solution at the end of this excerpt, the comparisons that Jace constructs shift his thinking in a notable way and mark an example of how he changes his mind.

> Dr. E presents $-5 - 4 =$ _____ on a piece of paper. Jace thinks silently for an extended time.
>
> *Dr. E: So, what are the thoughts going through your head?*
>
> *Jace: Like, I'm thinking maybe . . . I'll write it down, but I think that negative 5 minus 4 would equal negative 1. (Writes $-5 - 4 = -1$.) Because, just like the last problem, it's a negative number and a subtraction. But I don't know.*
>
> *Dr. E: What has you kind of questioning yourself right now?*
>
> *Jace: Because . . . ah, maybe not. You know what, I think it's 9 because if you have a negative 5 and you flip the problem around—so, 4 minus negative 5—that would be 9 because you are taking away a negative number from the 4 even though you don't have a negative number. So,*

it would be plus instead of minus a negative. I'll wait a second. (Writes 4 − −5 = 9.)

Dr. E: All right, so you first thought it was negative 1.

Jace: Mm-hmm.

Dr. E: And now you don't think it is anymore? What made you change your mind?

Jace: Because it's . . . to be negative 1, it would have to be negative 5 minus negative 4 [−5 − −4 = −1], because 5 minus 4 equals 1 [5 − 4 = 1], and then they're all negative numbers. But since it's not, then it's going to be a different answer.

Dr. E: OK. So then you wrote 4 minus negative 5 equals 9. (Points at 4 − −5 = 9.)

Jace: Mm-hmm.

Dr. E: How come you switched the order of the numbers? (Points at 4 − −5.)

Jace: Because I think it helped me understand it better.

In this vignette, Jace shared his thinking about −5 − 4 by making comparisons, which is consistent with what researchers find from students who are engaging with integers for the first time (Bishop, Lamb, Philipp, Whitacre, & Schappelle, 2016a, 2016b, 2018; Lamb, Bishop, Philipp, Whitacre, & Schappelle, 2018). Prior to this excerpt, Jace solved problem types like −5 − 4 by first computing 5 − 4 and then making the solution negative. He specifically did this strategy for problem types where the minuend has greater magnitude than the subtrahend (for example, |−5| > |4|). In this excerpt, Jace first solved −5 − 4 by comparing it to 5 − 4 and then making it negative, consistent with his incorrect strategies prior. However, in the middle of the vignette, Jace notably shifted away from this type of reasoning. He changed his mind and shared that −5 − 4 cannot equal −1 by comparing −5 − 4 to −5 − −4 = −1. He shared, "To be negative 1, it would have to be negative 5 minus negative 4, because 5 minus 4 equals 1, and then they're all negative numbers."

Although Jace did not obtain the correct solution to −5 − 4, he did change his solution from −1 to 9 when he also compared −5 − 4 to 4 − −5: "I think it's 9 because if you have a negative 5 and you flip the problem around . . . that would be 9." Jace, an elementary student engaging as a young mathematician, constructed his own way of thinking about integers and shifted his own thinking. Shifting thinking and changing minds may result in a correct solution, or it may be a slow process with which you have to be patient.

Facilitating opportunities for students to create and change their own reasoning is an important aspect of supporting instructional experiences that build on student thinking. To do this, consider asking students to pause their work on a task well before they are finished solving it. Ask students to share their strategies for approaching the problem with the class (notice you aren't focusing on the solution!). Then ask them, "After hearing how some of your peers are working on this problem, do you want to change your approach?"

In this excerpt, you not only saw Jace's thinking about integers but witnessed his *learning*, or changes in his thinking. We chose this excerpt because it showed Jace engaging in the creative act of mathematics as he changed his mind. Further, we shared this vignette to normalize the importance of changing minds without an emphasis on the correctness of solutions. Jace is a great example of how changing one's mind in mathematics may or may not be related to being correct. He also is a great example of creatively inventing strategies and being OK with changing one's mind. Although Jace was still incorrect in his determination of $-5 - 4 = 9$, he significantly changed his thinking.

PAUSE AND PONDER

Jace changed his thinking from $-5 - 4 = -1$ to $-5 - 4 = 9$. In both cases, he made mistakes and had an incorrect solution. In what ways was he learning? If Jace were your student, what would you do next?

Changing minds in the field of mathematics is so normalized that even the most brilliant and renowned mathematicians do it in their most creative acts. For example, remember the story of research mathematician Andrew Wiles and Fermat's Last Theorem from chapter 4 (page 69)? Wiles references changing his mind in what he has called "the most important moment of [his] working life" (as cited in Mozzochi, 2004). Through changing his mind and revising his work, Andrew Wiles is now credited with having proven one of the most famous unsolved problems, and has won numerous awards, including the Abel Prize, which is like the Nobel Prize for mathematics.

Sometimes, learners imagine solving a mathematics problem as a clean, quick process. However, mathematics problem solving is a messy process with productive struggle (Baker, Jessup, Jacobs, Empson, & Case, 2020; NCTM, 2014). In the vignette, we saw Jace, a young mathematician, do creative work that entailed changing his mind, just like Andrew Wiles did. In both cases, the mathematicians engaged in the slow process of thinking, created new things, and changed their minds.

Teachers need to change their minds and try new strategies in mathematics. Once they recognize that mathematics is a creative act, and includes changing minds, they can integrate changing minds as a pedagogical tool in the classroom.

CHANGING MINDS AS A PEDAGOGICAL TOOL

The next layer of the Changing Minds Task Structure is changing minds as a pedagogical tool in the classroom. In this layer, teachers develop pedagogical approaches that facilitate students' consideration of different strategies. Changing minds as a pedagogical tool supports and encourages students to change their minds during the act of doing mathematics. This layer stems from the prior layer in that teaching changes once you see changing minds as an act of creativity in mathematics. Additionally, this layer differs from changing minds as a task because teachers implement and consider practices that encourage students to change their minds rather than developing a specific task a priori. These practices could be planned before instruction or may need to be pulled in the moment during a discussion.

One of the Mathematical Practices recommended by the Common Core State Standards (NGA & CCSSO, 2010, p. 6) is that students "make sense of problems and persevere in solving them." Persevering in solving problems inherently assumes students will make mistakes. And, to encourage perseverance in problem solving, NCTM (2014) recommends that teachers "support productive struggle in learning mathematics" (p. 48), when students encounter a mathematical problem and the answer is not obvious. They may explore different solution strategies. They may play with different mathematical ideas (and thus make a mistake). They may not determine a solution the right way or may determine a wrong solution. They may change their mind, feel challenged, and even feel frustrated. Changing minds as a pedagogical tool honors this perseverance from students and ways that teachers support productive struggle.

Changing minds as a pedagogical tool is especially important when, during the act of doing mathematics, you notice that some or many students are either (1) not using an efficient strategy or (2) struggling unproductively to reach the correct answer.

In chapter 5 (page 88), we introduced using unknown student work as a strategy for presenting mistakes. In this chapter, we are presenting a new strategy about supporting changing minds. However, we are also using unknown student work that can encourage learners to change their minds as they are doing mathematics. We end the section with other suggestions for impromptu teacher moves that can encourage changing of minds.

In the following vignette, Ms. B facilitates a conversation among students and encourages them to change their minds about the way they determined the mean of a data set (0, 20, 20, 20, and 20) using unknown student work.

> *Ms. B: OK, so the question we had to explore was, "What is the mean of the set of numbers 0, 20, 20, 20, and 20?" Let's hear how Diamond thought about this problem, because she had a great way to think about this pictorially.*
>
> *Diamond: I thought about them as stacks of unit cubes. And I saw that the first stack had 0 unit cubes, and the other four stacks had 20 unit cubes. And I thought about how I could move some from the four stacks of 20 to the first stack to make them all the same size. But instead of taking 5 from the stacks of 20, I took 4 from each stack because then each stack would have 16 cubes and be the same height.*

> **Ms. B**, writing the solution and reasoning on the board: *OK, I see just what you were thinking, Diamond. Now, what I want to do is share with you how one of my previous students, Komal, thought about this problem. Komal told me that she also got 16, but she didn't do it the way Diamond did it. Komal said that 20 plus 20 plus 20 plus 20 was 80.* (Writes that on the board.) *And then if she thought about divvying up 80 dollars among five people, then giving 1 dollar away to each person until there were no more dollars left would be 16 dollars per person.* (Writes the reasoning on the board.) *I want these pairs of shoulder partners* (gesturing to half the class) *to take Diamond's reasoning and these pairs* (gesturing to the other half) *to take Komal's reasoning. In your pairs, tell me the answers to the following question* (showing this on chart paper): *"In what situations would this way of solving the problem be useful?" As you are working, it would be helpful to think about it as, "When might you want to change your mind about how to solve this problem and use Diamond's way, and when might you want to use Komal's way?"*

In this vignette, two correct solutions were provided; both Diamond (the student) and Komal (the unknown student) shared 16 as the solution. However, Diamond's approach to finding the mean was more efficient than Komal's in this particular situation. While Komal's solution ultimately yielded the correct answer, it takes significantly more time to divvy up 80 dollars 1 dollar at a time. However, if Ms. B encouraged Komal to repeatedly subtract a number larger than 1 dollar, her strategy could be as efficient as Diamond's. (You can create similar tasks by changing the divisor and dividend to be more suited either to Diamond's strategy or to Komal's strategy.)

In this situation, Ms. B encouraged students to think about changing their minds about a strategy rather than a solution. She facilitated discussion for the students to consider the context of the problem and the complexity of the numbers. In this way, she was developing procedural fluency on the foundation of conceptual fluency by having students think about the problem's structure and the ways in which they could find the mean. Because Ms. B purposefully facilitated discussion to share thinking and allowed for opportunities for minds to change, she drew on the layer of changing minds as a pedagogical tool. Figure 6.3 (page 112) shows Diamond's and Komal's strategies on an anchor chart that a teacher made for encouraging changing minds.

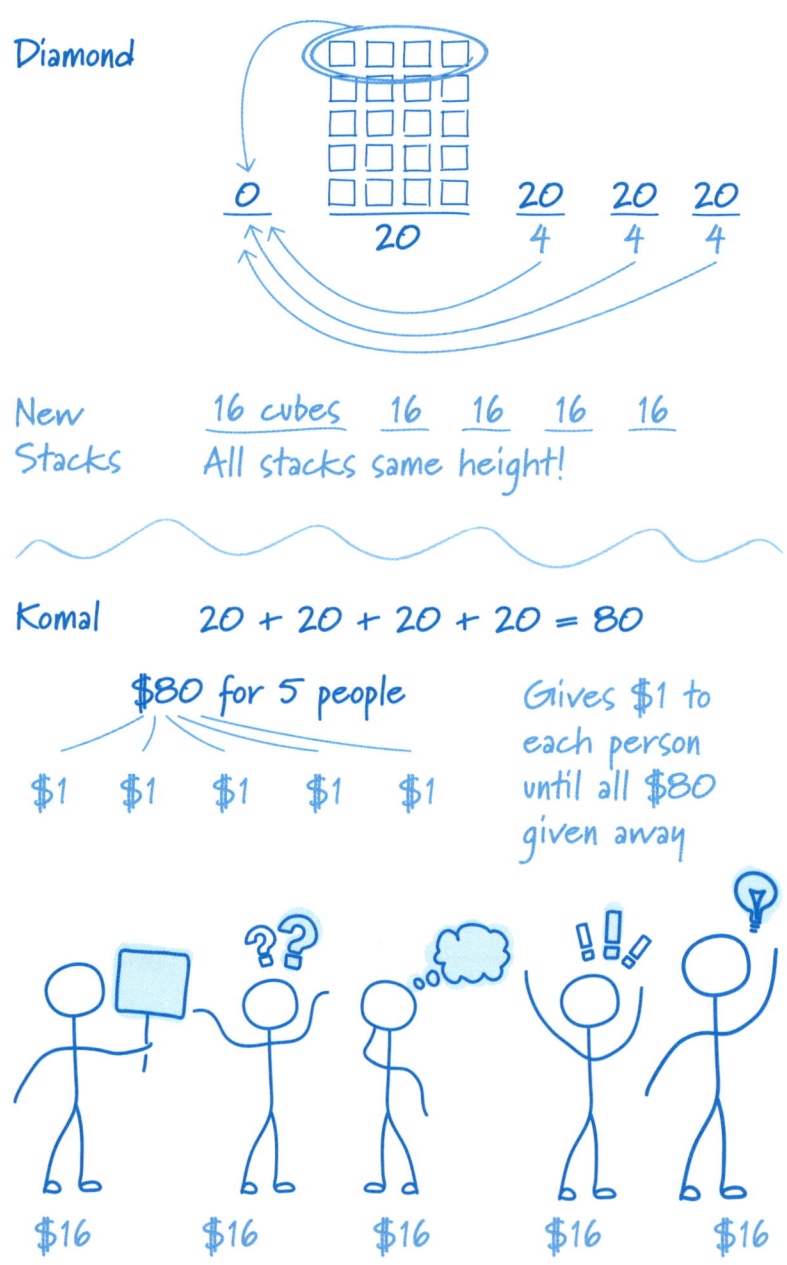

Figure 6.3: Diamond's and Komal's strategies.

Here, we illustrated how a teacher integrated unknown student work to encourage changing minds about how to find the mean for a data set. Changing minds as a pedagogical tool in the mathematics classroom is important for supporting and celebrating mistake making. First, an inherent attribute of valuing changing minds is valuing mistakes, like changing from an incorrect solution to a correct one. There is no better asset-based perspective than celebrating mistakes, and supporting changing minds is part of that. And second, to do mathematics deeply, one has to make mistakes. As one productively struggles and articulates ideas further, minds change.

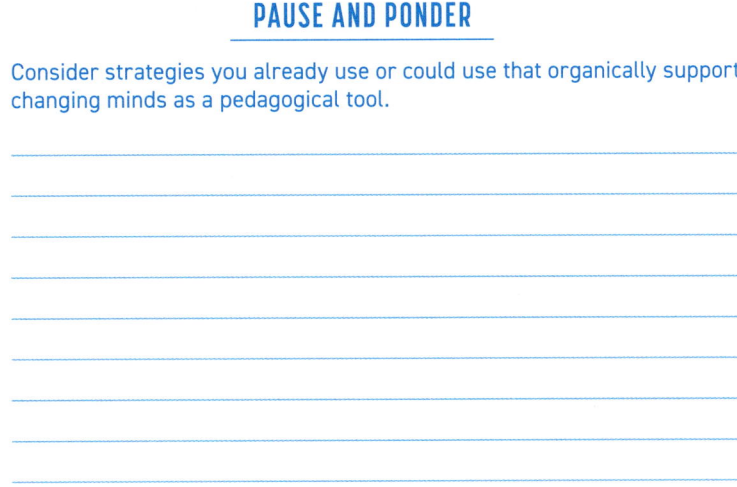

PAUSE AND PONDER

Consider strategies you already use or could use that organically support changing minds as a pedagogical tool.

_____
_____
_____
_____
_____
_____
_____
_____
_____

We have seen teachers support changing minds in their classrooms and how it has played out in our own. Some pedagogical strategies that are naturally supportive of changing students' minds as they engage in the act of doing mathematics are the following.

- Number talks
- Gallery walks
- Think-pair-shares
- Deep mathematical discussions

Number talks are a short (ten- to fifteen-minute) talk routine in mathematics where students share their thinking and strategies (Parrish, 2010). As part of the routine, students construct their own thinking, which requires creativity and definitely will include mathematical mistakes.

Number talks can be a space where students are supported to change their minds in community through discussion.

For gallery walks, students put their work, or the work of their group, up on the wall in an anchor chart like figure 6.3. Students then walk from poster to poster, exploring and making sense of others' thinking. The teacher can support this exploration in various ways; they can direct students to provide feedback to the poster authors (by writing it on a sticky note to place on the poster), to write their own ponderings on paper as they examine each poster, or to review all posters and then return to their poster and, if needed, revise their thinking. During this time, students will be exposed to different visual representations, new strategies, and even different solutions. Through making sense of others' ideas in a gallery walk, students have opportunities to change their minds.

The pedagogical strategy of think-pair-share can be used on its own in a mathematics lesson, but can also be incorporated within strategies like number talk. Think-pair-shares are where students think on their own and then share their thinking with a partner. Like number talks and gallery walks, think-pair-shares provide students opportunities to hear others' perspectives that are different from their own and that may change their minds. In our own classrooms, we often verbally make this idea of changing minds explicit during number talks and think-pair-shares. We will say, "Is anyone changing their mind about something? If so, why?" We will even follow up this question with a think-pair-share.

Deep mathematical discussions cover a rich mathematical task and are characterized by rich and vibrant discourse. These discussions may take longer than a few minutes on a task. They focus on conceptual understanding of (and mistakes about) a mathematical concept rather than on a procedure, the solution, or even the aesthetics of a problem (for example, "I like the colors she used for her representation!"). These are common in earlier grades or with students who are asked to focus on mathematics instead of aesthetics. If a deep mathematical discussion is being facilitated, changing minds will be invited and make the discourse richer.

One attribute these pedagogical strategies have in common is that they allow for multiple strategies from students. Additionally, all these pedagogical strategies give students the chance to solve exciting, interesting, or thought-provoking problems.

When Nicole visited her daughter's kindergarten classroom as a volunteer, she observed one of the kindergarten teachers, Mrs. N, using some of these pedagogical strategies as she supported mistake making and changing minds in mathematics. The next excerpt shows how Mrs. N supports the kindergartners in changing their minds as a pedagogical tool paired with think-pair-shares and deep mathematical discussion.

> The kindergartners gather in a circle, sitting on a carpet, with Mrs. N in the center. She has a large box of dried black beans and a second box of containers and pots.
>
> Mrs. N: Friends, can someone tell me what *capacity* means?
>
> Job: Is it a place?
>
> Mrs. N: It's something we use in mathematics for measuring.
>
> Mabel: Is it how much something is?
>
> Mrs. N: Yes, it's a way of measuring how much. For example (holding up a container), *capacity refers to how much I can fill inside of here*, like how much water could fit in here, how much rice could fit in, or how many of these beans could fit in. Friends, I am going to give you each two containers. With the person next to you, I want you to decide which one of these would hold more black beans when I fill them up.

Mrs. N passes out two containers to each pair of students. The containers are all unique. The kindergartners compare their two containers visually, guessing which one will hold more beans and why. Mrs. N encourages each of the pairs to talk it through. After supporting a think-pair-share, she then asks the students to share their initial thoughts with the group. She does not tell the students if they are right or wrong, only to share their thinking. Eleanor and Mabel have a tall, skinny cylinder and a short, wide cylinder.

> Mrs. N: Eleanor, will you share what you and Mabel discussed?
>
> Eleanor: I think the tall cylinder would hold the most because it's tall.
>
> Mrs. N: Thank you for sharing, Eleanor. Mabel, do you think that too?
>
> Mabel: No.
>
> Mrs. N: What do you think and why?
>
> Mabel: I think the shorter cylinder would hold more black beans because it's just a little shorter, but a lot wider.
>
> Mrs. N: Interesting thoughts! Thanks for sharing, Eleanor and Mabel. I wonder how we could figure out which one would hold more black beans.

Mrs. N asks each of the pairs to share and does not correct their thinking. Sharing takes a little longer than she anticipated. When it is time for recess, Mrs. N makes a pivot and adjusts her plan to continue the conversation.

> Mrs. N: All right, class, these are great thoughts. We will come back to these after you return from recess.

The students go to recess, and on their return, Mrs. N brings them back to the carpet. She asks the class if they thought more about their containers and, wanting to support deep mathematical discussion, asks, "How can we determine which of the containers holds more black beans?"

After getting ideas from the students, she supports them in making consistent scoops to fill the containers. She begins with the containers that Mabel and Eleanor considered. She has Mabel and Eleanor come to the center and fill their containers with black beans. The taller, skinnier container takes three and a little more scoops of black beans. And the shorter, wider cylinder takes four of the same-size scoops. After that demonstration, Mrs. N asks if anyone changed their mind. Eleanor shares, "I changed my mind now. The container that is shorter holds more black beans." Mrs. N reassures her, "It's OK to change our mind. It was hard to tell which one of these containers would hold more. We needed to investigate to know for sure."

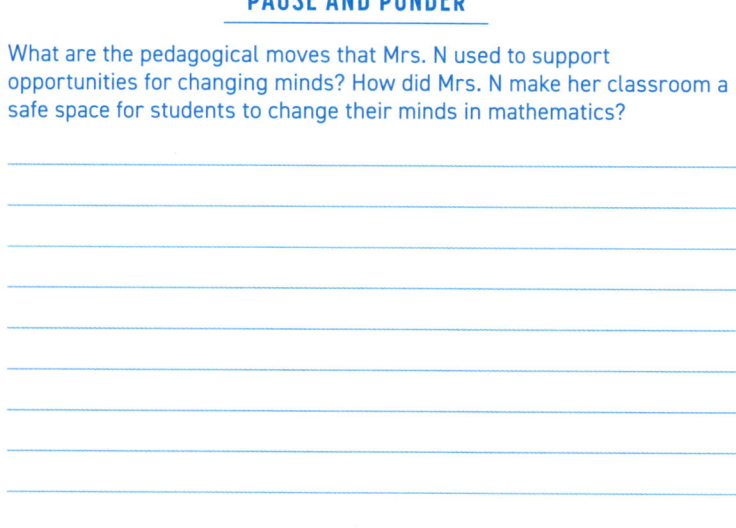

PAUSE AND PONDER

What are the pedagogical moves that Mrs. N used to support opportunities for changing minds? How did Mrs. N make her classroom a safe space for students to change their minds in mathematics?

CHANGING MINDS AS A TASK

The final layer of the Changing Minds Task Structure is changing minds as a task. A changing minds task highlights student work where a student is changing their mind about a strategy or solution, possibly from an incorrect to a correct solution. Basically, a changing minds task shows how a student, for a moment in time, was unsure or shifted their way of thinking in any way (incorrect to correct, correct to incorrect, incorrect to incorrect, or correct to correct with a focus on the latter correct method being more efficient). These tasks can highlight a piece of unknown student work and are selected prior to instruction (in contrast to the previous layer, where changing minds is encouraged in a more organic way within the mathematics activity). The piece of unknown student work could be something you create yourself based on your prior experiences or observations of teaching a mathematical topic. Or you may consider using student work you have collected over time or at a specific moment within your classroom. If your classroom is a space where student work is normally shared and discussed, you could also select pieces of work in action from your classroom.

Figure 6.4 shows an example of this task structure with a correct solution and an incorrect solution. Students are asked to examine how this student could have thought about both 3 and −1 as the solutions for the number sentence ___ − −2 = 1. This type of task honors that minds change in mathematics, and there is sense making behind each solution (both correct and incorrect).

☐ − −2 = 1

The student was changing their mind between 3 and −1 as the solution.

$\boxed{3}$ − −2 = 1
$\boxed{-1}$

Figure 6.4: An example of the Changing Minds Task Structure with student work.

When creating a changing minds task, you can use unknown student work or draw on work that you have collected within your own classroom. First, pick a topic and a specific task that you know students struggle with or have gotten incorrect in the past. If you are using unknown student work, solve the problem in two ways—using a correct strategy and a common strategy that results in an incorrect solution. If you are using student work from your classroom, listen to groups discussing a task and pick two strategies that are logical (although maybe not both correct). For example, when students engage with negative integers for the first time, inventing their own strategies, they grapple with subtracting a negative and sometimes solve problems like $a - -b$ by ignoring one of the minus signs, solving $a - b$ (Bofferding, 2010). Because of this, we chose a representative example of student work in figure 6.4 that captures that conflict (where the student crossed off a solution of 3 for ____ – –2 = 1, which they initially solved as 3 – 2 = 1).

The key to creating this task is that less is more. You do not want to create a long list of problems; rather, create one or two rich, discussion-worthy situations. For example, the work depicted in figure 6.4 also makes use of a number line. The student who solved this problem creatively invented a strategy with the number line, where –1 – –2 could be modeled by starting at –1 and moving right and losing 2 pieces of "negative distance" to 1 (see Wessman-Enzinger, 2019, for more about this way of thinking about subtracting a negative integer). In this case, the student work we selected reflects a changing mind about a solution and a visual representation to discuss as well. This one task can be used as mathematical discussion.

Using changing minds tasks offers three main benefits.

1. It *normalizes changing minds* about strategies and solutions.
2. It provides space for *critiquing the reasoning for all mathematics* (both correct and incorrect).
3. It supports *leveraging mathematical understandings about mistakes forward*.

With this task structure, we see the student in figure 6.4 changing their mind from 3 to –1, shifting from an incorrect to a correct solution. By making this shift in solution transparent in the task, we normalize the notion that students will change their minds. By capturing the changing mind from 3 to –1 with an accompanying number line, we then create a task that is open for critique. And finally, the goal is that students' learning moves forward. It is OK to be wrong and change one's mind, and this task structure captures the essence of learning. People change their minds about mistakes and, therefore, they can learn.

Concluding Remarks

How do you start with changing minds in mathematics? One way is to pick the layer that is most engaging or meaningful to you. For example, maybe you feel like you already recognize mathematics is a creative act that requires changing minds; then start at layer 3 with creating your own changing minds task. Or maybe learning about mathematics as something creative really rocked you. Then start at layer 2, brainstorming ways that supporting changing minds in mathematics would impact current structures in your classroom. The "Chapter 6: Changing Minds Task Structure" reproducible on page 121 can help with implementation.

Reflection Questions

Use the following questions for reflecting on the ideas in chapter 6.

1. Thinking about the excerpt featuring Jace early in the chapter (page 106), how can you support the slow and purposeful nature of changing minds? What can patience with changing minds look like in your classroom?

2. In what ways do you think changing minds in mathematics supports creativity in mathematics? Why do you think that?

3. What are some benefits of using changing minds as a pedagogical tool or as a task structure in your classroom?

4. Select a mathematical content area that interests you. Think of a common mathematical mistake in that area, and create a changing minds task for it.

Chapter 6 Application Guide

In this chapter, we explored the idea of changing minds to support mathematical mistakes in your classroom. Specifically, we introduced the layers of changing minds: a creative act, a pedagogical tool, and a task structure. The reproducible "Changing Minds Task Structure" (page 121) offers help for this task. Use the following application guide to connect the chapter's themes to your classroom.

Chapter Theme	Connection and Application to Your Practice
Changing minds as a creative act	Mathematics is creative! It is a creative act when students invent strategies in mathematics. Like anything creative, when students create their own strategies, they will make mistakes.
Changing minds as a pedagogical tool	You can embrace changing minds as a general pedagogical tool. What this means is if you value changing minds in mathematics, then you will support this in multiple spaces and treat your students as mathematicians.
Changing minds as a task structure	The Changing Minds Task Structure is a generalized structure for using unknown (or known) student work that illustrates a student changing their mind. The task invites other students to think about that student's thinking and describe why they are changing their mind or why they might want to change their mind. This task structure supports students' sense making, efficiency, and procedural fluency.

Chapter 6: Changing Minds Task Structure

Instructions: This structure has a student changing their mind between option A and option B as the solution. To this structure, add your own student work that shows a student is changing their mind. It does not matter if the student is changing their mind from a correct solution to an incorrect one, an incorrect solution to a correct one, or even an incorrect solution to another incorrect one. The purpose is to illustrate changing minds about a strategy and solution and get your students discussing it.

Option A: A student first thought . . . (Option A is a way for you to highlight a correct or incorrect strategy with a solution.)	Option B: A student changed their mind to . . . (Option B is a way for you to highlight a different correct or incorrect strategy with a solution.)
Add student work.	
Option A: A student first thought . . .	Option B: A student changed their mind to . . .
Have a student explain why they think option A or option B is correct.	
Describe what the student might be thinking.	

Celebrating Mathematical Mistakes © 2025 Solution Tree Press • SolutionTree.com
Visit **go.SolutionTree.com/mathematics** to download this free reproducible.

CHAPTER 7

This or That Tasks

The This or That Task Structure can be used to support the use of mistakes as a way of deepening conceptual understanding.

> Before reading this chapter, take a moment to reflect on or journal about the following questions: Have you paired a mathematical mistake in a discussion with a correct solution? How did the discussion go? What is an example of a mathematical mistake you and your students could reflect on within a mathematical discussion that enhances conceptual understanding?

I love anything that makes mathematics feel more fun, joyful, or playful. If mathematics is fun or joyful, students are more interested. They may also feel supported to take risks in trying new ways of thinking in mathematics. For these reasons, I always start class with a "Would you rather?"

question. Sometimes, these questions are mathematical—for example, "Would you rather buy jeans for $50 or buy jeans for 20 percent off?" What's fun about a "Would you rather?" question is that there is no right answer, and this can lead to lively debate and discussion. To make class more personal, I even use "Would you rather?" questions that are *not* mathematical. I do these every day for attendance—for example, "Would you rather have fur or scales?" "Would you rather jump on a cloud or a slide down a rainbow?"

These questions do not have right or wrong answers, which creates a safe space for talking about mathematics and learning more about each other as people. My students often tease me that I collect "Would you rather?" questions. I am definitely known as the "Would you rather?" teacher. But as Natasha and I were writing this book, we encountered people who questioned, "A book about mistakes? But there are *right answers*, right? So how do you help students realize that there is a right answer?"

★ ★ ★

It's true that there are right answers. My disclaimer is that sometimes there is a right solution, but in mathematics, sometimes there's no solution. Sometimes, there are unsolved problems that *might* have a solution no one knows yet. And there are mathematical problems that have multiple solutions. So helping students realize there is a right answer is a little complicated in a world where sometimes there is no right answer or more than one. Mathematics is not a field with perfectly distinct boundaries and categories where solutions are always either right or wrong. But I do agree, sometimes in mathematics there are right or correct solutions, especially in K–12 mathematics. And that is when the This or That Task Structure, which will be discussed in this chapter, really shines. The This or That Task Structure is perfect for teachers and students who like the fun nature of "Would you rather?" questions. Yet the This or That Task Structure is ideal for highlighting one correct solution and one incorrect solution (or any permutation you can think of). This chapter discusses the This or That Task Structure and explains how to create this or that tasks.

This or That Task Structure

As mathematics teachers, we are situated in a paradoxical loop: We love and value our students' mistakes as creative acts, and we want our students to develop strong conceptual understandings that lead to correct solutions. When we found ourselves within this paradox, we developed the This or That Task Structure to honor both correct and incorrect solutions simultaneously. NCTM (2014) recommends that teachers both build "procedural fluency from conceptual understanding" (p. 42) and "facilitate meaningful mathematical discourse" (p. 29). The This or That Task Structure achieves both! It is a tool for supporting mathematical discourse that deepens conceptual understanding and builds procedural fluency by discussing correct and incorrect solution strategies in tandem.

With the This or That Task Structure, we highlight one correct solution and one incorrect solution. One of them, this or that, is correct, and the other is incorrect. When we use the tasks that we present in this chapter, we share the following directions with our students: "We are going to show you students' solutions, drawings, and work. One of the solutions is correct, and one is incorrect—this or that. Help us!" Figure 7.1 shows a simple task we presented to students.

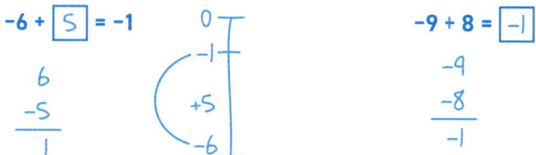

Figure 7.1: This or That Task Structure.

The structure of this task honors that we can, of course, be in pursuit of a correct answer while also valuing an incorrect solution. By intentionally unpacking both correct and incorrect strategies and solutions, we value both. Yet using the incorrect strategies and solutions deepens conceptual understanding about a topic by allowing students to learn from the mistakes. Thus, we can work toward a trajectory of correctness and also simultaneously value incorrect strategies and solutions.

How to Create This or That Tasks

When creating a this or that task, first pick a mathematical concept (for example, counting by tens, subtracting with regrouping, or finding the median of a data set). In figure 7.2 (page 126), we selected the topic of fraction comparison. We began with the task of having students answer, "What fraction is larger: 4/7 or 4/10?" From our experience implementing this task, we knew it would be good for eliciting student discussion, as it contains common numerators instead of denominators. And from our students' prior work on this topic, we knew the mistakes that come from this task are often conceptual in nature (see chapter 3, page 53) and

> **THIS OR THAT?**
> **What fraction is larger: 4/7 or 4/10?**
>
> "I think 4/7 is smaller than 4/10 because 7 is a smaller number than 10."
>
> "I think 4/7 is larger because it is bigger than a half, and 4/10 is less than a half."
>
> **Figure 7.2:** This or that fraction comparison task.

worth discussing. These support students in thinking about comparing fractions in a different way than common denominators.

Second, consider a mistake, strategy, or solution that will enhance mathematical discussion and deepen conceptual understanding. Because we knew that many students use common denominator for comparing fractions, and we wanted to support other strategies for thinking about fractions, this task was an ideal choice. We purposefully wanted to offer strategies in our this or that task that differed from common denominator in order to deepen students' conceptual understanding of fraction comparison.

> 💡 As you teach concepts, make note of mistakes your students make that are conceptual. Try to document how the student reasoned their mistake was correct and what helped them move toward a correct solution. These notes will help you in future iterations of teaching these concepts in a way that celebrates mistakes. 💡

You may choose to use actual student work as examples, perhaps work recently completed in class or work from a prior lesson. Or you may choose to draw on your knowledge to anticipate mistakes your students may make. Using known or unknown student work, select two different strategies or solutions—one that has a mistake and one that does not. One useful strategy that our students have invented to do tasks like the preceding fraction comparison task is using ½ as a benchmark fraction. When students use ½ as a benchmark fraction, they compare both 4/7 and 4/10 to ½. Another strategy we see students invent is considering the size of pieces (that is, comparing the size of sevenths to the size of tenths). For example, students will compare 1/7 and 1/10, stating that 1/7 pieces are bigger than 1/10 pieces because sevenths are partitioned into fewer pieces.

This or That Tasks

Yet by this time, students have learned a procedure for getting a common denominator to compare fractions and often use that. For that reason, we wanted the task to encourage students to use either comparison to the benchmark fraction ½ or the common numerator strategy, not the common denominator strategy. We thought those would be two strategies to include in a this or that task. As a result, students offered the idea that $4/7$ is greater than ½ because they verbally reasoned $3.5/7$ equals ½. We intentionally left out $3.5/7 = ½$ in figure 7.2 to offer the opportunity for discussion and addition. However, you could add that in. For the other strategy, we wanted to use a mistake that would lead to a discussion on conceptual, and not just procedural, aspects of fraction comparison. By pointing out 7 (the whole number) is smaller than 10 (the whole number, as shown on the left in figure 7.2), we drew on a common mistake that students make when comparing fractions. Although 7 is less than 10, $1/7$ is greater than $1/10$. Even if a student already knows that $1/7$ is greater than $1/10$, this type of mistake offers the student a chance to describe why $1/7$ is greater than $1/10$.

The two main benefits of integrating this or that tasks into your mathematical discussions are the following.

1. It *normalizes mathematical mistakes* as part of doing mathematics by making them discussion worthy.

2. It supports *using mathematical mistakes to support and deepen conceptual understanding* of mathematics.

Consider the this or that task in figure 7.3, where ___ – –5 = 0 is solved in two different ways (–5 on the left and +5 on the right). This task differs from the one presented in figure 7.1 (page 125) because it uses the same exact task answered in two different ways.

$\boxed{-5}$ – –5 = 0 $\boxed{5}$ – -5 = 0

-5
$\underline{-\ -5}$
0

Figure 7.3: This or that task for ___ – –5 = 0.

When we implement this or that tasks, our students sometimes think both items provided are incorrect, even though we tell them one is correct. In the following vignette, we highlight Hudson (a fifth grader engaging with negative integers for the first time) and how he reasons both solutions are incorrect. We think Hudson's assertion that both are incorrect would be a rich opportunity for opening debate and discussion in class.

> *Teacher: So, this says box minus negative 5 equals 0, and the student on the left wrote in negative 5, while the student on the right wrote in positive 5.*
>
> *Hudson: I don't know!*
>
> *Teacher: Can you tell us what you're thinking about?*
>
> *Hudson: I wanna agree with the one on the left but, like, they're both minus, and if it's negative, it's gonna . . . it's probably gonna equal negative, negative 10. And the one on the right, it's still minus, and I feel like it would equal. I don't know what the other one equals, but I feel like they're both wrong.*
>
> *Teacher: You feel like they're both wrong? Um, can you explain what you were saying about 10? I didn't understand that part.*
>
> *Hudson: Oh, um, I was saying negative 5 minus negative 5. I thought it would equal negative 10.*
>
> *Teacher: Oh, OK, thank you. Can you tell me why you think they're both wrong? You've explained why you think negative 5 minus negative 5 is wrong. Can you explain why you think the one on the right is wrong?*
>
> *Hudson: OK, I don't know if it's wrong, but I feel like it's not wrong, but I feel like it is wrong.*
>
> *Teacher: OK, can you tell us why maybe you think it might be right and tell us why you think it might be wrong? We'd love to hear that, how you feel unsure.*
>
> *Hudson: I think it's right because it's negative 5. I mean, positive 5, not negative 5, but then if you look at the other one, hmm . . . I don't know.*

This or that tasks also reveal students' conceptual understanding around a topic. In the following excerpt, Estrella, who completed the same this or that task for ____ – –5 = 0, identifies the correct and incorrect solutions incorrectly. Yet, by explaining the reasoning on the left, she describes that the solution is –5 and uses strong analogical reasoning by comparing –5 – –5 = 0 to 5 – 5 = 0. Although Estrella also thinks 5 – –5 = 0, she understands the reasoning for –5 – –5 = 0. Through the this or that task, we learn Estrella created her own term, "dual negatives," for subtracting a negative and, although she doesn't recognize –5 – –5 = 0 as the correct solution, she is developing conceptual reasoning that could support that as a solution.

Teacher: So, this says box minus negative 5 equals 0, and on the left, the student wrote in negative 5, and on the right, the student wrote in positive 5.

Estrella, pointing at the problem on the right: Let's go with that.

Teacher: Does that mean the one on the right?

Estrella: Yes.

Teacher: OK, tell me more.

Estrella: Because, like I said in the last one, 5 minus—take off that negative because it doesn't explain to subtract positive by two negatives, because then you could never get a positive number. And then you take off that negative, so 5 minus 5 equals 0.

Teacher: So, I heard you say that you could never get a negative when subtracting a positive? Is that what you said? Can you help me?

Estrella: No, I mean, like, you can never get a positive from two negatives in a row.

Teacher: How do you know?

Estrella: Because it would be, like, I don't know exactly, except I don't really know. What I've seen, I've never come across a dual negative that equals a positive.

Teacher: Can you explain what's challenging about the dual negative?

Estrella: Because, like, it's kinda confusing because every once in a while, you mix up the negative with the negative—the minus sign with the negative sign—so that you think it's like 5 minus 5, so it's like dual minuses. Every once in a while, you mess up like that.

Teacher: Can you explain more about what you think about the student on the right with the crossing off of the negative? Can you explain more about that?

Estrella: She's showing, like, because the negative 5, it's subtracting 5 from that, and she's taking out that negative because it's already a negative, so to make it less confusing, and then 5 minus 5 is 0.

Teacher: I have one more clarifying question about that. So originally, it said box minus negative 5 equals 0, but then the student did cross off that sign on the right. Do you think these are two of the same questions? Is it asking the same thing even when the student crossed it off?

Estrella: Yes.

Teacher: Thank you. And then for the problem on the left, tell us how you think this student got negative 5.

Estrella: Because he tried to match the thing so that it would be negative 5 minus negative 5, and like in positives, if you have two numbers that are the same, they take each other out. Except in negatives they just form together, I believe.

Teacher: That was a really helpful explanation for me. Thank you. So, if you were to solve this problem as shown, what do you think the answer would be?

Estrella: Five.

Teacher: Positive 5?

Estrella: Yes.

Teacher: Can you tell me more?

Estrella: Because if, like they said on the right, 5 minus 5 equals 0. So, they fight each other, and they tie. So, it turns into nothing.

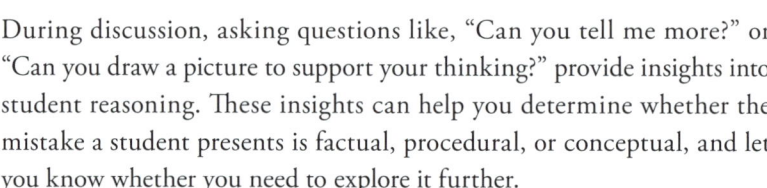

During discussion, asking questions like, "Can you tell me more?" or "Can you draw a picture to support your thinking?" provide insights into student reasoning. These insights can help you determine whether the mistake a student presents is factual, procedural, or conceptual, and let you know whether you need to explore it further.

Concluding Remarks

How do you use the This or That Task Structure? Any way you want! Use one correct strategy or solution and one incorrect strategy or solution. Use the same problem for both this and that or use two different problems that target the same idea. The example in figure 7.1 (page 125) used two different problems that target the same idea (subtracting a negative number). The examples in figures 7.2 and 7.3 (pages 126 and 127) used the same problems with two different solutions. Use actual student work (figures 7.1 and 7.3) or draw on knowledge you have curated about common mathematical mistakes from your students (figure 7.2). Start with this task structure as an introduction to a lesson, use this as a number talk, or use the discussion on this for your entire lesson. You can also use the reproducible "Chapter 7: This or That Task Structure" on page 133. Have fun discussing this or that with your students!

This or That Tasks

Reflection Questions

Use the following questions for reflecting on the ideas in chapter 7.

1. What are the benefits of using the This or That Task Structure in your classroom?

2. What ideas do you have about how you could implement this or that tasks in your classroom?

3. Select a mathematical content area that interests you. Think of a common mathematical mistake in that area, and create a this or that task for it.

4. What are ways, in addition to task structures like This or That, that you can normalize mistakes in your mathematics classroom?

Chapter 7 Application Guide

In this chapter, we explored the idea of the this or that task to support mathematical mistakes in your classroom. We highlighted the this or that task as a vibrant space for deepening conceptual understanding by using one strategy that has a mistake and one strategy that does not. The reproducible "This or That Task Structure" (page 133) offers help for this task. Use the following application guide to connect the chapter's themes to your classroom.

Chapter Theme	Connection and Application to Your Practice
The this or that task normalizes mathematical mistakes as part of doing mathematics by making them discussion worthy.	Keep track of mistakes your students make that are discussion worthy to use in a future this or that task. As you track these, consider questions you might ask your students to help them analyze the mistakes and move them toward your learning goal for the activity.
The this or that task supports using mathematical mistakes to support and deepen conceptual understanding of mathematics.	When using the this or that task, your goal is to deepen your students' conceptual understanding of a topic. Mistakes you choose to highlight should therefore not simply be factual or procedural but conceptual, as we discussed in chapter 3.

Chapter 7: This or That Task Structure

Instructions: The essence of this task structure is to create a scenario of two different points of view, preferably where one has a mistake in it. Add your own student work to this structure.

This? Highlight a correct strategy with a solution.	That? Highlight an incorrect strategy with a solution.

Ask students, "Which did you pick—*this* or *that*—and why did you pick it? Why didn't you pick the other option?"

CHAPTER 8

Invented Notation and Language

BIG IDEA

Supporting invented mathematical notation and language is a way to support creativity that will naturally result in mistake making. Yet some of the mistakes that happen with supporting invented notation and language in the mathematics classroom are not really mistakes at all.

> Before reading this chapter, take a moment to reflect on or journal about the following question: What is something as a mathematical learner you were told was "wrong" but is not really wrong?

A REFLECTION FROM NICOLE

I remember writing down the expression 2 – 7 on my paper when I was in second grade. I have a vibrant memory of my teacher, Mrs. R, exuberantly telling me, "No, no, we always write the bigger number first and the smaller number second." She even turned my mistake (or was it?) into a whole-class discussion. Much to my surprise later on, my fifth-grade teacher, Mrs. H, introduced the idea of negative numbers to my class, and I realized that 2 − 7 is indeed just as legitimate as 7 − 2. My mistake in second grade was not a mistake in fifth grade when Mrs. H introduced the existence of negative numbers. Therefore, the rule that Mrs. R told my second-grade class came with an expiration date (Karp, Bush, & Dougherty, 2014).

In addition to negative numbers, Mrs. H introduced the idea of square roots to my class. I remember being fascinated by square numbers and the idea of taking the square root of them and other numbers. I was playing with the ideas of square numbers and square roots on my calculator during some free time. Using the square root key on my scientific calculator, I tried to take the square root of a negative integer (since I now knew negative numbers existed). An error message appeared on my display, and I asked Mrs. H, "What is this error?" She replied, "You can only ever take square roots of *positive* numbers." Like my second-grade teacher, Mrs. H used this as a teaching opportunity for the whole class. We all discussed how it is nonsensical to take the square root of a negative number. Well, maybe you can see where this is going. In high school, my tenth-grade mathematics teacher, Mrs. K, introduced the complex number system, or the amazing world of imaginary numbers. I learned that I could take the square root of negative numbers. Again, taking the square root of a negative number was not really a mistake. These are all examples of rules or procedures that expired when I learned more mathematics (Karp et al., 2014).

I wish I could say these stories of "mistakes that are not really mistakes" ended here. But that is not the case at all. The stories followed me into university mathematics as a mathematics major and into my own teaching life. For example, Mrs. K went on to teach me, "We can never divide by zero," and I learned in calculus and real analysis that weird things actually happen when you divide by zero (not that you "can't do it"). I also learned in high school that triangles "always have angles that sum up to 180 degrees." I later learned in graduate school this is true in Euclidean geometry, but a spherical triangle can have angles that sum up to more than 180 degrees. The takeaway here is many of my mistakes in school mathematics were not *really* mistakes, and thus, a goal of this chapter is to discuss how some mathematical mistakes are not really mistakes at all. Maybe some things you or your students are doing in the mathematics classroom look like mistakes but instead are creative and can even extend learning opportunities in your space. "Mistakes that are not really mistakes" especially happen when you allow students to invent mathematics (notation, language, and strategies) themselves like young mathematicians.

★ ★ ★

In this chapter, we unpack how students can invent their own notations and language within mathematics and how these inventions, although sometimes viewed as incorrect, are actually to be celebrated and encouraged in the classroom. Mistakes are proof your students are trying. And perhaps mistakes are also evidence of new ways of thinking that may not be mistakes.

Invented Notation and Language

When teachers support their students as mathematicians and provide space for them to engage in creating mathematics, their invented notations and language emerge. Using invented notation and language promotes mathematical learning because it supports students' contributions and prior knowledge and welcomes mistake making. NCTM's (2014) recommendations highlight the importance of using and connecting mathematical representations. Inviting students to share their thinking means also welcoming their unique perspectives, which may differ from traditional notation or representations. Of course, you can (and should) connect their representations to traditionally used ones. And you should also simultaneously celebrate students' unique and unexpected contributions.

When you support invented notation and language, you will notice that some "mathematical mistakes" are not really incorrect at all. Instead, they are instances where students have created robust, sophisticated mathematics that society has said is incorrect because it is not what is traditionally used by mathematicians, mathematics curriculum, and other instructional materials within school mathematics. What follows are various examples of invented notation and language that may seem incorrect but are actually really sophisticated invented mathematics.

ZERO NUMBERS

In our work with third and fourth graders, we solved problems and played games where we supported students in creating their own notation. For example, when solving problems or playing games, we gave students number paths that did not have all numbers filled in. Rather than giving the third and fourth graders a number path with negative numbers on it, we

celebrating mathematical mistakes

provided the students with a number path that had blanks, as seen in figure 8.1.

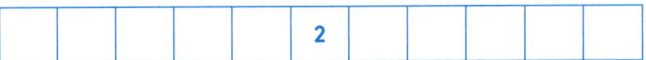

Figure 8.1: A number path with only a positive integer.

When we did this, students became creative about what numbers could go on either side of the 2. For example, in figure 8.2, we include two number paths that Mario (a fourth grader) created.

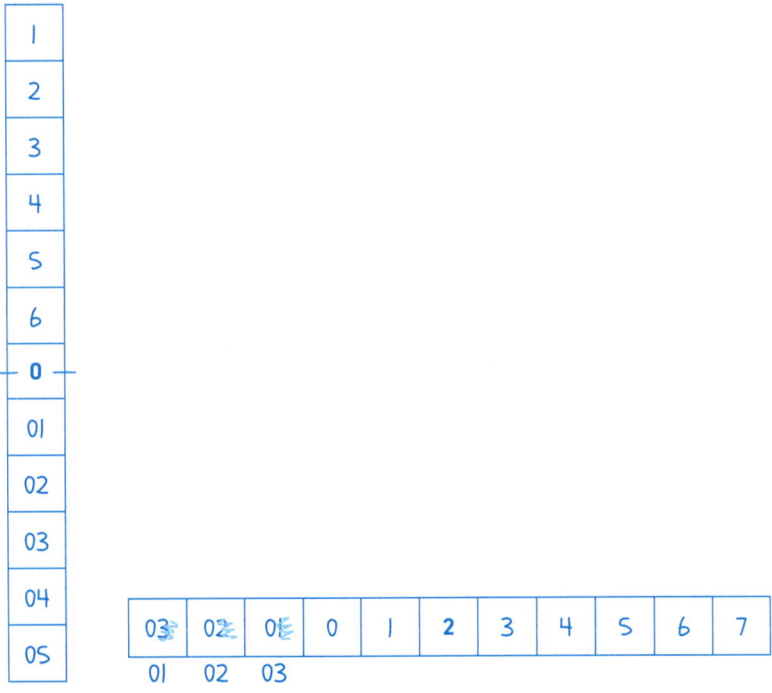

Figure 8.2: Mario's zero numbers.

In figure 8.2, you may notice that Mario was changing his mind about the order of the numbers, or you may find a mistake in how Mario ordered positive integers in the vertical number path (1, 2, 3, 4, 5, 6, 0). However, we also hope you notice the brave creativity that Mario shared by creating a new type of number. He created the idea of a *zero number*.

Invented Notation and Language

PAUSE AND PONDER

What do you think about Mario's notation or zero numbers? Is this a mistake or something else?

_____
_____
_____
_____
_____
_____
_____
_____
_____
_____

Mario, and other students, invented the notation of 01 for representing –1. Although 01 is not –1, the third and fourth graders treated 01 as –1. In fact, the young mathematicians even used this notation when solving problems like 5 + ___ = 3. For 5 + ___ = 3, they determined that the solution should be 02 (or –2). We do not consider Mario's zero numbers a mistake. Rather, we think of 02 as a sophisticated mathematical invention that is just like –2 but different from what is traditionally used in mathematics. Students, especially young mathematicians, should be supported in their invented notation and language before formal academic language and notation are introduced. Supporting students with invented notation and language begins with facilitating opportunities for students to think creatively about mathematics by opening spaces for inventing or reinventing mathematical ideas. Further, when students invent notation and language, it is essential that you do not tell the students they are "wrong" when their invented notation looks different from what you see in your curricular materials (for example, zero numbers).

For supporting students' invented notation and language, provide them with tasks that have more than one strategy for solving them or have multiple solutions. This can allow for their creative notation and language to emerge.

celebrating mathematical mistakes

NEGATIVE NUMBERS ON THE RIGHT-HAND SIDE

Figure 8.3 highlights how a fifth grader, Jace, drew a nontraditional number line for solving –6 + ___ = 15. In this drawing, –6 is on the right-hand side of his number line. Traditionally, –6 appears on the left-hand side of a horizontal number line. Although this is different in that it does not align with what is taught in traditional mathematics, it is not incorrect. As you can see, the fifth grader used this number line to correct thinking about moving 21 units from –6 to 15.

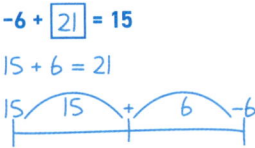

Figure 8.3: Negative integers on the right-hand side of an empty number line.

Jace shared the following as he solved –6 + ___ = 15.

> *Jace: I did 15 plus 6 because the answer . . . Since 15 is a whole number, and then that would be just regular 15, but you have to add 6 more because the 6 goes . . . Hold on. Here, I will draw you one. (Draws the number line in figure 8.3.) So that would be negative 6 right here (draws the negative number on the right) and 15 right here (draws the positive number on the left; 15 and –6 are each an equal distance from 0 in the drawing). It would be 15 (draws an arc from 15 to 0 and writes 15 above the arc) plus (draws a plus sign above the 0) another 6 (draws an arc from 0 to 6 with 6 above the arc). Ah, negative . . . wait, 21. (Writes 21 in the box.) Just regular 21.*

PAUSE AND PONDER

What do you notice about Jace's number line and his strategy for solving –6 + ___ = 15?

_____
_____
_____
_____
_____

Invented Notation and Language

We notice that Jace said "regular numbers" instead of "positive numbers." Yet this did not interfere with his creating a robust empty number line and obtaining a correct solution. We also notice that Jace placed positive numbers (or "regular numbers") on the left-hand side of the number line and negative integers on the right-hand side. Although that is not conventionally done in mathematics, Jace still produced a valid, sophisticated, and correct strategy that led to a correct solution—and as mathematics teachers, we should celebrate! In fact, positive and negative numbers are relative numbers and can be placed like this on the number line, although it is not traditionally done.

What's best for supporting a young mathematician like Jace who is thinking about and creating mathematical ideas for the first time? Is it a good idea to stop his flow and ideas to say, "They are not regular numbers," or "Do not put negative numbers on the right side of the empty number line"? Rather, we think it is important to encourage such thinking. We think what Jace did is brilliant. We also know Jace did not make mistakes; rather, he invented new ideas and communicated creative ways to expand the whole number system.

When you are presented with invented phrases (like "regular numbers") or notation like we see in Jace's work, we encourage you to allow this creativity. As learning progresses, we can introduce the mathematically acceptable notation and language in a way that respects Jace's work by saying something like, "Remember how Jace introduced us all to his number line strategy and 'regular numbers'? Well, in mathematics, we often refer to these numbers as *positive integers* and then the negative numbers as *negative integers*." To respectfully move Jace's number line representation to a more mathematically acceptable notation, consider showing the traditional representation on the board as unknown student work alongside Jace's work. Then ask students what they notice and wonder about the two different representations. Encourage them to identify ways they are similar and different. End this discussion with acknowledgment that although Jace's representation is correct, the other representation is more often experienced and examined within the curriculum and day-to-day mathematics work.

Encouraging your students to draw representations to model their work without telling them what to draw (or similarly asking them to use manipulatives to model their work without telling them which manipulatives to use) encourages creativity and invented notations.

SAYING NUMBERS USING *AND*

Miss Sailer, a second-grade teacher, shared with us a mistake she saw when co-teaching. In her own second-grade classroom, Miss Sailer and her co-teacher introduced three-digit numbers to their students. Some of the second graders engaged with three-digit numbers for the first time. When first encountering three-digit numbers, the students needed to say (or invent) what number they saw. One of the second graders looked at the number 107 and said, "One hundred and seven." When seeing 219 for the first time, the second grader said, "Two hundred and nineteen." Miss Sailer's co-teacher critiqued this language as incorrect because the student used the word *and* between "two hundred" and "nineteen." This co-teacher argued that the student's language was incorrect because *and* is reserved for decimal places. She critiqued this language outside of class, sharing that the student should have said, "One hundred seven" and "Two hundred nineteen," instead.

PAUSE AND PONDER

What do you think about the *and* language here in verbally describing three-digit numbers?

_____
_____
_____
_____
_____
_____
_____
_____
_____
_____

Miss Sailer reflected on the student's use of *and*. She disagreed with her co-teacher because she saw that use of the word connected to visual representations of the number as seen in figure 8.4 (page 143). In this figure, the hundreds flat and the ones cubes are separate entities. Miss Sailer shared how if you visually show 107 with base 10 blocks, for instance, you can see one flat of 100 *and* seven small cubes each representing 1, where there is 100 *and* 7 ones.

Invented Notation and Language

Figure 8.4: The number 107 represented with base 10 blocks.

When Miss Sailer first shared this story with us, we thought about how this use of *and* is really similar to our students' placement of negative integers on a number line's right-hand side in that these actions are typically identified as mistakes. But are these *really* mistakes? Or, do we just tell students they are wrong because society has told us there is only one way to say numbers or write numbers? Are we stifling our students' creativity and mathematical prowess by labeling their invented notation and language as incorrect when it simply does not fit within mainstream mathematics? These types of invented notation and language prompted us to reflect on our own practice. We began to consider ways to support invented language and notation more.

PAUSE AND PONDER

Like Miss Sailer, have you encountered something in your own practice that you thought was not really a mistake but others told you was incorrect?

_____
_____
_____
_____
_____
_____
_____
_____

DIFFERENT BASES (THE CASE OF BASE 5)

In this section, we will explore a mathematical task that we think supports invented notation and language, and naturally invites mistake making. But first, let's use our imaginations to consider a mathematical idea that you likely do not use in your mathematics classroom. It's going to feel strange, but we want you to persevere with us because we know this task supports mistake making and enhances numerical reasoning. Get ready to imagine!

Here's the base 5 task: What if our ancestors decided to count with only one hand instead of both hands? Perhaps this seems like a weird question. But this is the scenario we want you to pause and ponder with us. We want you to use your imagination to think about all the things that would change if our ancestors did things differently.

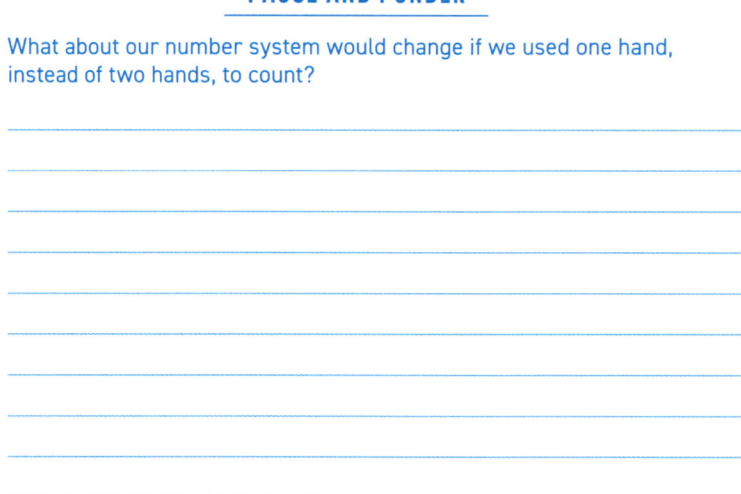

PAUSE AND PONDER

What about our number system would change if we used one hand, instead of two hands, to count?

_____
_____
_____
_____
_____
_____
_____
_____
_____
_____
_____

If our ancestors had decided to count and group only by the fingers and thumb on one hand, what would our place value system be? Let's start with digits. All our numbers are made up of digits. And in our system, where we use two hands instead of one hand, there are ten digits (0, 1, 2, 3, 4, 5, 6, 7, 8, and 9). And those ten digits are in positions, or place values, that give value to the numbers. For example, consider the number 123. The 1 digit is in the hundreds place, giving a value of 100 to the number; the 2 digit is in the tens place, giving a value of 20 to the number; and the 3 is in the ones place, giving a value of 3 to the number.

Invented Notation and Language

I hope you are noticing at this point that it is not a coincidence we have ten fingers and ten digits. So, if we used one hand instead of two hands to count, then we would have only five digits. But what would those digits be? When we asked our students this question, a lot of them answered as follows: 1, 2, 3, 4, and 5. That's definitely on the right track, but we still need a 0 in our place value system, and including that digit with 1, 2, 3, 4, and 5 would make six digits.

There are a lot of ways to justify this other than the inclusion of zero, but let's keep going.

Given that we can't use the digits 1 through 5 if we want 0, our digits with one hand are 0, 1, 2, 3, and 4—one hand, five digits. What kinds of numbers can we make with our five digits? Just like working with ten digits, we can make an infinite amount of numbers using these five.

Let's take a look at these numbers. Here's what some base 5 numbers look like: 14_{five}, 23_{five}, 144_{five}, 312_{five}, 20.3_{five}. We have placed *five* in subscript to the right of the numbers to indicate our numerals in the new counting system we are thinking about. But what do these numbers mean? Do they mean the same thing in this new number system? And what would counting look like?

If we used one hand instead of two hands, what would come next: 1, 2, 3, 4, . . . ? It feels like 5, but remember, we do not have a 5 digit anymore. Let's consider an analogy with base 10. Our current counting system with two hands and ten digits is called a *base 10 system*. We count 1, 2, 3, 4, 5, 6, 7, 8, 9, . . . and then 10. The 10 is made of two digits, 1 and 0, with the 1 digit in the tens place and the 0 digit in the ones place. What happens here is the 10 ones group together and make a ten with no ones.

This extends across our base 10 system: 10 tens make 100, 10 hundreds make 1,000, and so on. This analogy can help us when thinking about using only one hand. Using one hand and five digits is a system called *base 5*. We start counting 1, 2, 3, 4, . . . and then we add our fifth block, but we do not have a 5 digit anymore in our new system. It becomes 10 base 5. How is this different from our current number system?

Well, the 5 ones come together and create 5, which means in our number system with one hand, we need a new place value system. And we need a new fives place (analogous to the tens place in our base 10 counting system). Then, we can put 5 fives together and make a 25, so our next place value would be the twenty-fives. Figure 8.5 (page 146) illustrates how linking cubes, grid paper, or base 5 manipulative pieces could be used to start to create a base 5 place value system.

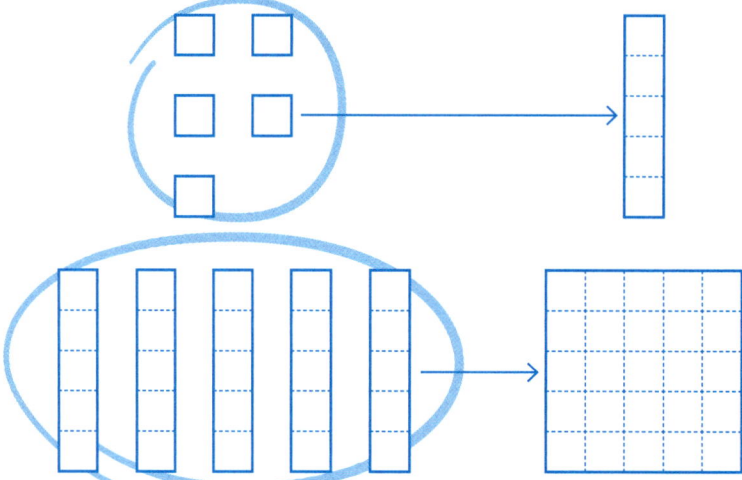

Figure 8.5: Grouping linking cubes, grid paper, or base 5 manipulatives in fives to build new pieces.

Figure 8.5 connects to the place value system in figure 8.6 because the ones circled represent ones (0–4) that can be used in the ones place. Once there are 5 ones, these can be made into a rod of 5, which is related to the fives place.

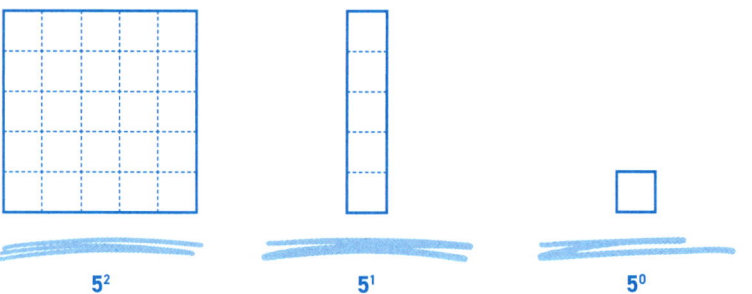

5^2 5^1 5^0

Figure 8.6: Creating a new place value system for base 5.

What we are doing here is creating a new place value system. Instead of ones, tens, and hundreds, we have ones, fives, and twenty-fives. Maybe this seems really strange. But this quinary (or base 5) system is related to the Mayans' numeral system, which ended about AD 900. Although they used a vigesimal (or base 20) system, we can see the influence of five in their system.

Counting systems have changed and evolved over time. In fact, they are still evolving; yet we are firmly positioned in our base 10 world.

Base 10 is convenient and, in fact, very efficient. But is base 10 an efficient counting system for computers and other machines that don't have ten fingers? We can rudimentarily think of our computers as machines with on-off switches (or circuits), and think of those switches as fingers. In that sense, I hope it makes sense our computers would profit from a base 2 system, or a binary system of numbers.

Have you ever done arithmetic in a base other than base 10—for example, base 5? Because students likely have no experience with different bases, it's a ripe mathematical topic for supporting invented notation and language.

Next, we will share some practical guidance for incorporating different bases into your classroom if you choose to try this topic, which we think is a different topic for ages 9 to 99. First, provide your students with linking cubes, grid paper, and markers for them to construct their own base 5 manipulatives, and encourage them to create a new place value system with the base 5 manipulatives they make. The intent is for students to have creativity in developing their base 5 manipulatives without the teacher directing them. Second, be patient as your students struggle with creating a new number system—this took centuries for mathematicians, so your students will not do this quickly. You can scaffold their development of a new number system by asking them questions like the following.

- How many digits would your new number system have? Why would it have that many?

- What stays the same in moving from the base 10 place value system to a base 5 place value system? What changes?

- How can you create manipulatives for base 5? In what ways are these similar to or different from base 10 manipulatives?

Just setting up a new place value system or creating manipulatives in base 5 could be a task on its own. Once this is done, you can have fun counting forward and backward in a different base system. Or you can have fun with simple addition problems, like $2_{\text{five}} + 4_{\text{five}}$.

How to Use Invented Notation and Support Invented Language in the Classroom

Next, we will share some ways that you can support invented notation and language in the classroom. These strategies include creating space and time for student invention of notation, listening to and pondering students' ideas, and using tasks that support mathematical imagination.

CREATE SPACE AND TIME FOR STUDENT INVENTION OF NOTATION, LANGUAGE, AND STRATEGIES

The first step for supporting student invention of notation, language, and strategies is simple: create space for it. When you desire to center student-invented notation, language, and strategies, all your pedagogical decisions focus on creating space for invented notation, language, and strategies. For example, rather than starting a lesson with a visual representation, like a number line or even the base 5 (or 10) manipulatives, you can ask students to draw a number line or create their own representations first. Additionally, rather than telling students which manipulatives to use for a task, ask them which manipulatives they think might help them solve the task at hand. Giving them choice with representations and tools provides a rich foundation for creativity. By centering invented notation, language, and strategies and creating space for them, you also welcome mistake making.

BRAVELY LISTEN TO AND PONDER STUDENTS' IDEAS

It can be intimidating to make sense of your students' ideas. For example, when I (Nicole) gave my students the base 5 task described previously, one student created his own symbols for base 5 instead of using traditional numerals. At first, I did not understand this explanation. However, I leaned into pondering how he could have come up with those symbols and supporting his conception of base 5. Figure 8.7 shows this student's invented notation for the base 5 system. He described the eye as 0, the vertical line as 1, the two lines as 2, the three lines as 3, and the square as 4. His invented notation is isomorphic to the digits 0, 1, 2, 3, and 4.

Figure 8.7: Student-invented notation for base 5.

USE TASKS THAT SUPPORT IMAGINATION OF MATHEMATICAL TOPICS

Some tasks support mathematical imagination better than others. For example, to support imagining negative integers, I (Nicole) created tasks that evoke the use of negative integers but do not explicitly contain negative integers (Wessman-Enzinger, 2023). The figures in this section show examples of these types of tasks. Figure 8.8 (page XX) highlights number sentences that use only positive integers as numerals but promote creating negative integers.

Invented Notation and Language

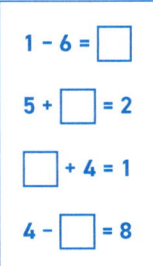

Figure 8.8: Number sentences without negative integers.

In addition, figure 8.9 and figure 8.10 are a number path and a number line with only one number. In both of these figures, students could decide not to use negative integers or to place them in unconventional spaces.

Figure 8.9: Number path without negative integers.

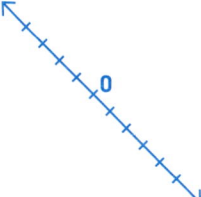

Figure 8.10: Number line without negative integers.

Concluding Remarks

What if instead of trying to correct students' mistakes, teachers celebrated their brilliant creations? Of course, you want your students to eventually learn accepted academic notation and language. But starting with invented academic notation and language allows you to begin with students' contributions, build on what they know, and welcome mistake making. In fact, sometimes what the world calls a mistake is just a different perspective of how a mathematical system could have been thought about. Use the "Chapter 8: Invented Notation and Language" reproducible on page 152 for ideas of open-ended tasks.

Reflection Questions

Use the following questions for reflecting on the ideas in chapter 8.

1. What is something in your classroom that you previously thought was incorrect but that might be correct? Why might it be correct?

2. Describe the types of spaces in your classroom where you could support the invention of notation and language.

3. Consider a mathematical visual representation you use to support students' mathematical learning (for example, fraction bars, number lines, or graphs). What is a way you can support students' invention or reinvention of that visual representation?

4. Among your experiences teaching mathematics, what is something your students created that stands out to you (either invented notation, language, or visual representations)? Why does it stand out? How could you use this in a future mathematics lesson?

Chapter 8 Application Guide

In this chapter, we discussed ways to support invented notation and language to embrace mistakes. Representations for the different tasks, such as the number paths for supporting invented notation, appear in the reproducible "Invented Notation and Language" (page 152). Use the following application guide to connect the chapter's themes to your classroom.

Chapter Theme	Connection and Application to Your Practice
Invented notation	Invented notation is when a student creates their own way to write a mathematical idea. This may or may not align with accepted academic notation, and you should stay open to the unique ways in which students invent notation. For example, in this chapter, we illustrated how a student wrote *01* instead of *–1* as their invented notation for a negative integer.
Invented language	Invented language is when a student creates their own way to represent a mathematical idea verbally. This may or may not align with accepted academic notation. For example, in this chapter, we illustrated how a student used the word *and* in ways that a teacher did not when stating numbers. Though this student's language differed from the teacher's, it did match the visual representation. Again, consider ways in which students' invented language appeals to the conceptual underpinnings of an idea. Although it may be academically incorrect, some invented language makes sense or is used in the real world.

Chapter 8: Invented Notation and Language

When you use open-ended tasks such as the ones featured in this reproducible, students may invent their own notation for negative integers (or they may make patterns). You can decide how to scaffold the discussion or modify the instructions to fit your students and classroom. The representations (number paths and number line) could be implemented all together or individually. We have found when we implement them all together, students are encouraged to think about various representations and how they can communicate similar or different ideas about order and magnitude.

Number Paths and Number Line

Instructions: Place numbers in the paths or on the number line.

HORIZONTAL NUMBER PATHS

				2					

				0					

VERTICAL NUMBER PATHS

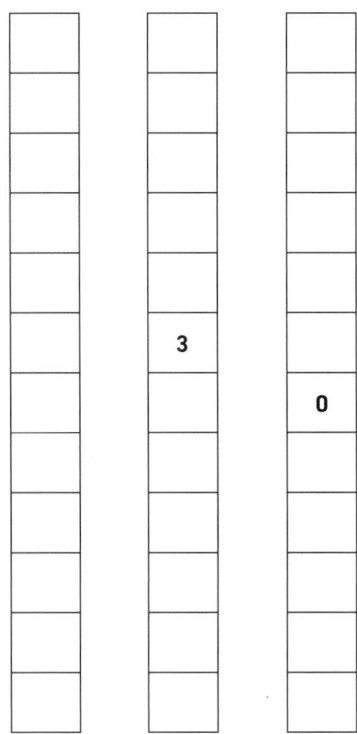

SLANTED NUMBER LINE

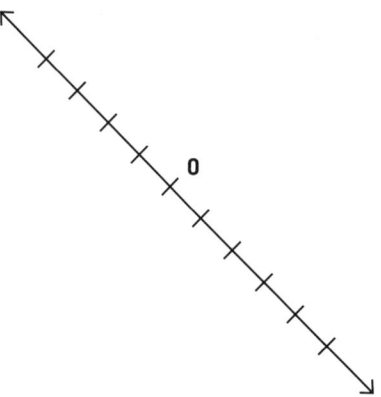

Number Sentences That Support Invented Notation

Instructions: Determine what number goes in each box. Write your solution in the box, and explain your reasoning.

1 − 6 = ☐

5 + ☐ = 2

☐ + 4 = 1

4 − ☐ = 8

CHAPTER 9

Mathematical Games

BIG IDEA

Using authentic play in mathematics goes beyond just using mathematical games. Mathematical games are not necessarily real play. Rather, intentionally integrating playful experiences (for example, games with tenets of play, and intellectual play) at all grade bands helps create a space where students feel safe to make mistakes to learn from.

> Before reading this chapter, take a moment to reflect on or journal about the following questions: How do you incorporate play into your mathematics lessons? In what ways do you think play can support and celebrate mistake making?

A REFLECTION FROM NATASHA

I have two young children who love to play board and card games as well as engage in imaginative play. My son's current obsession is the number infinity. He asks me questions like, "How big is infinity? Can we count to infinity? What is infinity plus 1?" Right now, he believes "infinity times a thousand" is bigger than "infinity plus one." As a mathematician and educator, I find it hard not to chime in with a brief lecture on the different types of infinities that exist and how we could prove whether "infinity times a thousand" is greater than, less than, or equal to "infinity plus one." But then I remember he's five. Five-year-olds love to play with ideas and especially love to play with numbers. He tells me his creative and playful ideas about infinity because he feels safe to do so. This sense of safety allows him to stretch his intellectual play muscles, and I am in no way concerned if he comes up with incorrect ideas about infinity. However, I am worried about developing a space where he and his sister feel encouraged to be creative in their thinking about numbers without worrying about whether their ideas are wrong. We spend a lot of time talking about numbers he can easily model and think about ("What is 3 plus 2, and is it the same as 2 plus 3?"), and we often do that through gameplay. (We love the game *Tiny Polka Dot*!)

As you read this chapter, think about how your students and even any other children in your life grapple with mathematical ideas and how this creative play develops curiosity and tenacity. Encourage gameplay and creative play in your mathematics classroom to inspire your mathematicians to study more mathematics! (I personally can't wait to hear if my son one day determines whether "infinity times a thousand" is greater than, less than, or equal to "infinity plus one.")

Mathematical Play

At the heart of mathematical anxiety is the fear of failing or being wrong. To counter this, we need to incorporate more joy and play around and within mathematics. A central tenet of the notion of play is that making mistakes is expected. Think about the first time you tried to ride a bike or play a new board game as a kid—you likely expected to fail. You may have gotten frustrated, but did not see failing as a sign you were incapable of doing that activity. Children naturally accept this tenet of their play. As such, incorporating play into mathematics makes sense as a foundation on which to investigate mistakes because it creates a space where mathematical mistakes are expected and are an inherent way to experience and do mathematics. In this chapter, we will discuss how to incorporate mathematical games into your classroom. But first, we need to elaborate on what we mean by *mathematical play* and distinguish how it relates to gameplay.

Mathematical play can be intellectual play and not necessarily gameplay (Featherstone, 2000; Wessman-Enzinger, 2018). And gameplay is not necessarily mathematical play. Researcher Herbert Ginsburg (2006) offers that through play, students deeply engage in mathematics, reminiscent of mathematicians:

> Young children develop mathematical strategies, grapple with important mathematical ideas, use mathematics in their play, and play with mathematics. Young children often enjoy their mathematical work and play. Indeed, despite its immaturity, young children's mathematics bears some resemblance to research mathematicians' activity. Both young children and mathematicians ask and think about deep questions, invent solutions, apply mathematics to solve real problems, and play with mathematics. (p. 158)

The idea of fusing play with mathematics is coming at a pivotal time in education and society. Increased educational testing (Ravitch, 2010), demands to meet educational standards (NGA & CCSSO, 2010), and needs for students to pursue STEM careers (Ellis, Fosdick, & Rasmussen, 2016; Olson & Riordan, 2012) are just some contemporary pressures. As stress has continued to build around testing and standards, there has also been a push to extend play, generally speaking, throughout elementary school (Parks, 2015). Incorporating play within mathematics may reduce stressful mathematical experiences, thereby decreasing students' opportunities to develop mathematical anxiety while also increasing their opportunities to play. Engaging students in playful experiences as mathematicians could also increase their access to complex mathematical concepts. Although most calls for mathematical play (Ginsburg, 2006; Wessman-Enzinger, 2018) and prolonged play (Parks, 2015) describe these play experiences as important for young students, we argue that mathematics learners of *all* ages should experience complex concepts through play. Here, we will illuminate mathematical play's and gameplay's potential for supporting all students' mathematical thinking and learning.

Mathematical play, like play in general, is a place where students can play with ideas in imaginative ways. These imaginative ways may be correct or incorrect, but in play, correctness does not matter. For example, if a child is pretending to fly or be a dragon, this is obviously not real (that is, correct), but it is real in play. In the same way, students can play with mathematical ideas—focusing on ideas and strategies that delight them without fear of whether they are real or correct. Table 9.1 (page 158) provides the tenets of play in general, which you can incorporate into mathematical play and games.

Table 9.1: Tenets of Play

Tenets for Play (Burghardt, 2011)	Additional Tenets for Play (Parks, 2015)
Spontaneous or pleasurable	Opportunities for social engagement
Not fully functional	Creative thinking
Different from similar serious behaviors	Appealing materials
Repeated	Physical movement
Initiated in the absence of stress	Imagination

A mathematical game is not necessarily playful on its own (Wessman-Enzinger, 2018). Yet students can draw on the tenets of authentic play as illustrated in table 9.1 when they engage in a mathematical game or task. When they do that, they are playing with numbers or mathematics. Jo Boaler (2016) supports play with numbers for all students: "The best and most important start we can give our students is to encourage them to play with numbers and shapes, thinking about what patterns and ideas they can see" (p. 34). Helen Featherstone (2000) argues that as children engage with integers, they may be in the "territory for mathematically imaginative play" (p. 20). She proposes that the integers themselves are this imaginative world, which is partly why you see them pop up throughout this book. Integers are a ripe place for all elementary and middle school students to engage in play. Featherstone (2000) writes, "The territory below zero is a separate world for elementary students. It is an outside the 'real' world of natural numbers—numbers that are in daily use both inside and outside of school" (p. 20). This type of imaginative play may be a way to share integers sooner and prolong play in schools. But you could include this play with other topics as well. Table 9.1 highlights general criteria of play to consider as you plan any mathematical task, but especially as you use mathematical games.

When considering how to incorporate mathematical play into your instruction, look for opportunities where using mathematical manipulatives (whether tactile or digital) can support the development of an idea.

Furthermore, we posit that mathematical play makes sense at almost every stage of learning, such as when you first introduce an idea, when students are learning strategies, when students are practicing, and at every age. Next, we will share how elements of play in table 9.1 can be incorporated into mathematical gameplay as described by Gordon M. Burghardt (2011) and Amy Noelle Parks (2015). Then, we will connect these instances of play (integer play and playing with integers) to the work of research mathematicians to show the potential for play in upper elementary grades and beyond.

Before my students began a mathematics card game I (Nicole) had planned, I explained the game's directions. I then asked the students who should go first. The following excerpt illustrates them engaging in play with integers in this setting. Their interactions represent authentic play as it went off course (in a good way) from the game I had planned (for more information about this, see Wessman-Enzinger, 2018).

Dr. E: So, I was thinking, how do we decide who goes first?

Kim: Rock paper scissors.

Alice: Or who draws the highest card.

Kim: Yeah, draw highest card.

Jace: Yeah.

Dr. E: OK, so everyone takes . . .

Jace: Everyone takes one card, and whoever has the highest . . .

(Alice, Jace, and Kim draw cards. Alice draws a −4 card, Jace draws a −8 card, and Kim draws a −7 card.)

Kim: I totally lost.

Alice: I did too.

Jace: I got negative 8.

Alice: I got negative 4.

Dr. E: OK. And you got what?

Kim: Negative 7.

Jace, pointing at Alice with −4: So she goes first.

Kim, pointing at Jace with −8: So Jace's is the highest, actually.

Alice: No, I am.

Jace: No, well . . .

Dr. E: So who's the highest?

Alice, raising her card in the air: Me!

Kim: Jace, because his is the biggest in the negatives. Because we all have negatives, so . . .

Alice: Well, mine would be the biggest.

Jace, pointing to Alice*: Well, she's the closest to one.*

Dr. E: So somebody said that they think Jace's is the biggest because it's negative 8.

Alice, shaking her head*: No.*

Dr. E: So why did you think that Jace's is the biggest?

Kim: I don't know. They're all negative numbers and just, like, find out which one is bigger.

Jace, gasping*: I was wondering why you would want to discard cards. I'm, like, if they are all whatever, why would you want to put one down? OK, now I see.*

Kim: Now I know why. (Holds the −8 card up in the air.)

Dr. E: And what's yours?

Alice: Negative 4. (Holds up her card.)

Dr. E: So which one do you think is bigger?

Alice: Mine.

Dr. E: Why do you think yours is bigger?

Alice: It's closest to 1. It's highest out of all of them.

Kim: Well, yeah.

Jace: Mm-hmm.

Kim: So I'm second. I'm second. (Waves her hands and card in the air.)

The students then drew two new cards and started the game. Although they never explicitly verbalized who should go first, Alice played first.

This excerpt highlights elements of play: function, creativity, social engagement, and the absence of stress. Alice, Jace, and Kim's suggestion of how to decide who should go first illustrated a *functional* component of playfulness (Burghardt, 2011); the students wanted to play the game and needed to decide who should go first, resulting in this mathematical discussion. This excerpt is playful because the students illustrated *creative thinking* (Parks, 2015); they created the ideas of order and magnitude when comparing integers. This excerpt is also playful because the students participated in *social engagement* (Parks, 2015); although they did not verbalize a conclusive agreement on which card was "biggest," they decided to let Alice go first and played without conflict. They freely had this discussion in the *absence of stress* (Burghardt, 2011); they decided how they would determine who would go first in excitement to begin gameplay.

Distinguishing between the order and magnitude of integers is an important component of understanding integers and represents prerequisite knowledge for integer addition and subtraction (Bofferding, 2014). Through deciding who should play the game first, Alice, Jace, and Kim played with integers as they initiated a discussion about order and magnitude. They found themselves grappling with order versus magnitude to compare their three negative integers (–4, –7, and –8). During this freely chosen activity, the cards revealed all negative integers serendipitously. Kim stated that –8 was "bigger" than the other numbers because –8 was "more negative"—employing magnitude-based reasoning (Bofferding, 2014). Alice and Jace reasoned that –4 was "highest" and "biggest" because it was closest to 1—employing order-based reasoning (Bofferding, 2014). Language issues of "bigger" and "higher" are also important tenets of the prerequisite knowledge that students need to make sense of as they begin learning addition and subtraction (Bofferding & Hoffman, 2015).

Traditionally, school mathematics emphasizes order over magnitude with integer comparisons. That is, when comparing numbers like –4, –7, and –8, –4 > –8 is expected because of order; –4 is close to zero on a number line, or –4 is farther right on a number line than –8. However, often the work of mathematicians is magnitude based. That is, there are times when –8 is "bigger" than –4. For example, consider two velocity vectors, one with magnitude –8 and another with magnitude –4. The vector with magnitude –8 would be considered "bigger." Also, this excerpt illustrated students engaging in play that became an unresolved mathematical problem for them around order and magnitude. Sometimes, mathematicians work on problems that are not resolved right away. This is the expected and normative work of mathematicians.

Gameplay

Games are opportunities for structured play that follow rules or use strategies with the end goal of someone "winning" or achieving some goal. Games that allow for strategic play, such as chess, can often help with developing conceptual mathematical thinking. Moreover, gameplay is an obvious way of incorporating playfulness into mathematics. As such, games that are naturally mathematical (for example, *Qwixx*, *Tiny Polka Dot*, and *SET*) are excellent tools to incorporate into your practice to develop students' willingness to make and learn from mistakes. However, we also want to mention that many games we see used in mathematics classrooms do not support mistake making in the way we have described it in this book. For example, *Around the World* and similar games that focus on speed and accuracy teach students that doing mathematics well requires them to be fast and correct. This is counter to our entire message.

Rather than detailing what games exist that are mathematical, in this chapter, we want to instead focus on the general implementation of such games to develop your practice. Furthermore, we want to highlight how mistakes, a natural part of playing games, lead to the synthesis of learning.

Prior to implementing games into your classroom routines, you have a few things to consider. First, as you ponder games for your classroom, look for games that develop understanding of concepts related to your students' needs or gaps. For example, if you are a kindergarten teacher and notice many of your students are struggling to subitize, you may use a game like *Tiny Polka Dot* because it can help develop this understanding. *Tiny Polka Dot* is a card game that features small cards with six different representations of numbers 0–10: (1) numerals, (2) dots in arrays, (3) scattered dots of varying sizes, (4) dots in a circle, (5) dots configured like the dots on dice, and (6) scattered dots of different colors but the same size. Game players have a variety of options with how to use the cards (all provided with simple instructions for even the youngest learners). A simple way to use the game is to use it for memory—and the parent or teacher can decide which cards to use to best support students (that is, which representations students need the most support with in understanding how to subitize). It is a playful game intended for learners ages 3 to 8. (We provide lists of games and resources we have found useful for getting started with gameplay in the reproducible "Chapter 9: Resources for Mathematical Gameplay" on page 171).

Areas where your students are struggling with procedural fluency are great places to support learning via games. Additionally, if you have students who need more prior grade-level content, particularly skills, consider incorporating gameplay on those skills, as it will encourage them to develop the skills in a low-stakes, highly engaging way.

Second, consider choosing games that allow for a variety of reading levels and backgrounds. For example, *SET* is a great game for readers of all levels because the cards do not contain words. This contrasts with a game like *Monopoly*, where reading cards and spaces on the board might be cumbersome and thus not joyful for those still developing their

Mathematical Games

reading skills. (For some ideas for games, see the reproducible "Chapter 9: Resources for Mathematical Gameplay" on page 171.)

Finally, consider how and when you will implement the game time in your class. Will it be part of daily instruction (some games allow for quick play), or should gameplay happen once a week? Will it make sense for students to play a variety of games within the class, or should you focus the entire class on playing one game (which would require enough game boards and pieces for each group of two to four to have a set)? Furthermore, we encourage you to consider always beginning gameplay in a collaborative way. By having students play collaboratively in a two-versus-two setting, you prepare the gameplay to include opportunities for discussion among pairs or small groups, which can solidify learning. Collaboration and discussion within gameplay not only supports mathematics but can create spaces of joy for students as well (Wessman-Enzinger & Bofferding, 2023).

Implement gameplay into your classroom early in the school year, and as your students play, remind them of your classroom norms and discussion norms. Games are a great location for students to practice these skills and develop their mathematics community.

Solidifying the basics of how and when you will implement the gameplay prior to implementation allows for a smooth transition to incorporating this into your practice.

Once you have chosen your game and have solidified the implementation details, let's unpack how you can implement gameplay in a way that develops conceptual understanding and highlights the play elements described in table 9.1 (page 158). When you do this, gameplay can become a space that requires mathematical mistake making in safe and playful ways. We have identified five steps for how to implement gameplay that can develop students' mathematical understandings.

STEP 1: IDENTIFY MATHEMATICAL GOALS

Identify the mathematical goals for students to learn in the game. This is a good idea for all tasks. NCTM (2014) states to "encourage mathematical goals to focus learning" (p. 12), and this is true for mathematical play and gameplay as well. If needed, adjust the game rules, board, or pieces to support the goals. For example, let's assume you implement fifteen minutes of game time once a week in your kindergarten class, and this week, you are going to play *Tiny Polka Dot* because students are learning how to subitize. However, you also know pockets of your students are struggling with identifying numerals and others are struggling with identifying quantities of dots that are not in arrays. As such, the learning goals for this gameplay will be the following.

- Develop subitization skills with quantities that are not in arrays.
- Identify numerals 1–10.

Since identifying quantities in arrays is not a struggle for your students, you are going to remove the cards that have quantities in arrays, and you are going to play a matching game with the remaining cards. Students will match numerals to their quantity, and they must justify their thinking to their partner.

Do not be afraid to adjust mathematical games to meet your students' needs. This can be as simple as adjusting the rules or the difficulty of the numbers in the game.

By identifying the mathematical learning goals, you are prioritizing the development of skills in a playful way that supports mathematics instruction. An administrator or parent would see this activity not simply as playing games but as *supplementing* regular mathematics instruction. Furthermore, identifying the mathematical goals supports the next step, which is identifying the game.

STEP 2: IDENTIFY THE GAME

To meet the learning goals established in step 1, you need a game that develops and supports those learning goals and has features that allow students to develop their identities as mathematicians. So, how do you decide what is a good mathematical game and whether it aligns with the learning goals? Some features of a good mathematical game include the following.

- **Does not emphasize speed:** Games that encourage speed (for example, *Around the World* and *War*) send a very clear message to players that if they are too slow, they are not good enough for this game. This runs counter to the message that we and many mathematicians have been trying to instill in our students—mathematical learning takes time, and to be "slow" simply means you are thinking deeply. Mathematicians hardly work on their problems quickly; many take years to solve a singular problem. Games that encourage speed negate all this hard work.

- **Relies on a strategy to solve:** Games that rely on strategy encourage players to develop a plan of attack. Players have to consider various things in a game of strategy, such as "Does it matter if I am player 1 or player 2?" and "How does my opponent's move hinder or support my next move?" Further, these types of games often have players considering an offensive approach versus a defensive approach to play. In the classroom, strategy games are wonderfully implemented in a collaborative play session—students who are working on the same team can develop their plan of attack together, further enhancing their communication and problem-solving skills.

- **Develops procedural fluency or conceptual understanding:** Games should develop procedural fluency and conceptual understanding. Games that develop procedural fluency are low-stakes ways for students to practice their skills. Using these games is a great tactic for students who are behind grade level or for those who might need some additional confidence in their skills. Games that develop conceptual understanding encourage students to use representations or reasoning to make their gameplay decisions, and these types of games often have a strong reliance on strategy.

PAUSE AND PONDER

What are games that you can think of that develop procedural fluency? Conceptual understanding?

_____
_____
_____
_____
_____
_____
_____
_____
_____

STEP 3: IDENTIFY FOCUSING QUESTIONS

Two types of questions you can ask students are (1) funneling questions and (2) focusing questions. *Funneling questions* "funnel" your students toward a predetermined strategy or solution. They are not about developing students' ideas or strategies. They are about a desired end result. These questions are not necessarily bad; they have their place, but using funneling questions takes away from play. Play involves creative and imaginative thinking, and funneling questions counter these tenets of play because they direct (that is, funnel) the student toward the answer without the opportunity for creativity or mistake making.

Instead, during gameplay, you want to use *focusing questions*, which researchers Beth A. Herbel-Eisenmann and M. Lynn Breyfogle (2005) describe as questions where, by listening carefully, teachers support students in sharing their individualized ways of thinking. By supporting individualized ways of thinking, teachers invite in creative and imaginative thinking, which is playful. We encourage you to examine Herbel-Eisenmann, Michael D. Steele, and Michelle Cirillo's (2013) set of discourse moves that can promote focusing questions. These discourse moves include waiting, revoicing, asking students to revoice another student's thinking, probing for student thinking, and encouraging students to engage with others' thinking.

Returning to our scenario of implementing *Tiny Polka Dot* and the mathematical goals identified in step 1 (page 163), you can come up with a list of focusing questions prior to gameplay. That way, as students are playing and you are circulating and observing their play, you are ready with questions and don't have to think of these on the fly. Focusing questions should develop students' communication skills as they pertain to the learning goals. So, for this example, focusing questions might include the following.

- How did you know that this numeral represents this quantity?
- How did you see this quantity in your head?
- Is there another way we could see this quantity? How so?
- Can you explain to me why you matched this numeral with this quantity? (Ask this if a student makes a mistake.)
- Why did you change your thinking? (Ask this if a student makes a mistake and then corrects it.)
- How do you know that student A is correct (or incorrect)? (Ask this of the partner who was listening to the justification.)

- Can you draw another representation of this quantity in a way that we have not seen yet? How do you know that your picture represents the quantity? Do you all agree? Why? (Ask the last two questions of the group.)

Although there are certainly many other questions that could be asked, this is a great starting list to support students as you monitor gameplay.

> Develop focusing questions that ask students to explain their thinking; justify their responses with pictures, words, or symbols; push them to develop other ways of thinking; and help them understand their own and others' thinking.

STEP 4: MONITOR IMPLEMENTATION AND PREPARE FOR SYNTHESIS

The fourth step is to monitor implementation of the gameplay. During this time, students are playing their games in their collaborative partnerships (two versus two). You will be circulating through the room, asking your questions as needed, and making note of what students are doing. Although this time may seem like an opportunity to step back, as a facilitator who is looking to connect their gameplay to learning goals, you must consider the students' play and determine how you can synthesize their work to solidify learning.

As students are working, make note of their mistakes that lead to aha moments. Ultimately, this is a way to approach formative assessment during mathematical play. Using and recording observational data can inform your decisions about subsequent teaching and learning. For example, if a pair of students incorrectly subitize a circle of nine dots, but through the focusing questions and work with peers, they see it as three groups of three, write that down! Consider repositioning the students' thinking for whole-class discussion. Use a discourse move to support further learning. Perhaps ask students to make sense of the pair's thinking or have one student revoice their thinking. This mistake leads to a key understanding for subitizing—flexibility in "seeing" quantities. Also, by writing it down, you are building a repository of student thinking that you can use later or use to inform teaching. Additionally, as you are monitoring, celebrate students when their mistakes lead to these moments.

STEP 5: SYNTHESIZE LEARNING

The most important part of any learning is synthesis. This gives students an opportunity to connect what they did in gameplay, particularly their mistakes, to the mathematical learning goals for the day. However, prior to this synthesis, give your students time to share their answers to the following questions with their neighbors and then with the class.

- What did you enjoy most about this game?
- What did you learn while playing this game?
- What will you do differently next time you play this game?

PAUSE AND PONDER

What other questions could you ask students to help synthesize their learning?

_____
_____
_____
_____
_____
_____
_____
_____
_____
_____
_____

After students have time to unpack their gameplay, share some of the mistakes students made that led to understanding. This presentation of student work can include student presentation and teacher presentation. It does not have to take a tremendous amount of time (taking as few as three to five minutes), but it warrants its own time because the power of synthesizing and presenting mistakes leads to learning. Sometimes, we like to have the synthesis at the end of a game take about a quarter of our time, with the bulk of our time spent on the game itself. If we notice that more discussion is necessary, we open our next lesson by centering on the context of the game, but devote the entire lesson to discussion. When we do this, students can see how their experiences and mistakes in the game enhanced their knowledge around the learning goals.

By taking gameplay through these five steps, with whatever game you choose, you are establishing the mathematical importance of gameplay by anchoring the entire session in the learning goals. Additionally, by putting such careful consideration into implementing gameplay, you are providing a safe and productive space for students to develop fluency through mistakes and joyful play.

Concluding Remarks

Although many are familiar with the importance of play in general and in mathematical learning in particular, in this chapter, we drew attention to how play in tandem with carefully selected games can highlight and leverage students' mistakes in a way that deepens their understanding. For games that are authentically playful, mistakes are expected. As such, they are a natural playground for students to make mistakes in and to learn from those mistakes.

Reflection Questions

Use the following questions for reflecting on the ideas in chapter 9.

1. How do games you play in your classroom align with the tenets of play in table 9.1 (page 158)?

2. How might you alter games you already play so that they meet those tenets of play?

3. What are times during your week when you can incorporate mathematical gameplay? Do you have an intervention time where gameplay makes sense or a community-building time in which this could fit?

4. In what ways can play and gameplay support mistake making in mathematics classrooms?

Chapter 9 Application Guide

In this chapter, we explored how mathematical games and play can support mathematical mistakes in your classroom while simultaneously developing procedural fluency and conceptual understanding. Use the following application guide to connect the chapter's themes to your classroom.

Chapter Theme	Connection and Application to Your Practice
Mathematical play is creative, inventive, and a natural part of being a mathematician.	Encourage your students to be creative and inventive by giving them time to play games that are not time based nor accuracy based.
Mathematical games are not a time for goofing off, but are highly structured, align with learning goals, and develop student learning.	If you take each of the five steps listed in this chapter, mathematical games can be seen as more than "just play." These games could even become part of your regular mathematics lessons if time permits, as going through these five steps can allow you to justify how the games fit into your scope and sequence.

Chapter 9: Resources for Mathematical Gameplay

The following lists suggest games that teachers can use to promote mathematical play.

Mathematical Games

EARLY CHILDHOOD GAMES

- Chutes and Ladders
- Dominoes
- Hi Ho! Cherry-O
- Jenga
- Tiny Polka Dot
- Too Many Monkeys
- UNO

ELEMENTARY SCHOOL GAMES

- ONO 99
- Qwixx
- Shut the Box
- Trouble

MIDDLE SCHOOL GAMES

- Prime Climb
- Skyjo

HIGH SCHOOL GAMES

- Catan
- Qwirkle
- Qwixx

ANY-AGE GAMES

- Blokus
- Chess
- Rummikub
- SET
- Yahtzee

Mathematical Game Books

ELEMENTARY SCHOOL

- *Mastering Basic Math Skills: Games for Kindergarten Through Second Grade* by Bonnie Adama Britt (2014)
- *Mastering Basic Math Skills: Games for Third Through Fifth Grade* by Bonnie Adama Britt (2015)

- *Math Fact Fluency: 60+ Games and Assessment Tools to Support Learning and Retention* by Jennifer Bay-Williams and Gina Kling (2019)
- *Math Games With Bad Drawings: 75 ¼ Simple, Challenging, Go-Anywhere Games—and Why They Matter* by Ben Orlin (2022)
- *Well Played: Building Mathematical Thinking Through Number Games and Puzzles, Grades K–2* by Linda Dacey, Karen Gartland, and Jayne Bamford Lynch (2015a)
- *Well Played: Building Mathematical Thinking Through Number Games and Puzzles, Grades 3–5* by Linda Dacey, Karen Gartland, and Jayne Bamford Lynch (2015b)

MIDDLE SCHOOL

- *Well Played: Building Mathematical Thinking Through Number and Algebraic Games and Puzzles, Grades 6–8* by Linda Dacey, Karen Gartland, and Jayne Bamford Lynch (2024)

Online Resources

- **NCTM Illuminations** (https://illuminations.nctm.org): Some of our NCTM Illuminations favorites include "Frustration: Analyzing a Card Game With Probability" (grades 6–8), "Factor Game" (grades 3–8), "KenKen for Younger Learners" (preK–2), "Polygon Capture" (grades 6–8), "The Game of SKUNK" (grades 6–8), and "Sticks and Stones" (grades 3–8).
- **NRICH** (https://nrich.maths.org/teachers): Some of our NRICH favorites include "Board Block Challenge" (ages 7 to 11), "Match the Matches" (ages 7 to 11), "Quadrilaterals Game" (ages 11 to 14), "Seeing Parallelograms" (ages 7 to 11), "Shapely Pairs" (ages 11 to 14), and "Totality" (ages 5 to 11).

CHAPTER 10

Mistakes in Action

There are a variety of organic and planned ways you can use the ideas from this book in a mathematics class to support mistake making in asset-based ways.

> Before reading this chapter, take a moment to reflect on or journal about the following questions: What is something about mathematical mistakes and their use in the classroom that you want to remember from this book? How are you going to celebrate mathematical mistakes in your classroom?

As with the previous chapters, we start this final chapter with a reflection, but in this case, we wanted to include a final reflection from each of us. We hope you have found these insights into our lives as mathematicians and mathematics educators inspiring and revealing. Additionally, we hope they encourage you to reflect on your history as a mathematician and mathematics educator.

A REFLECTION FROM NICOLE

One of my students joked with me in front of our class, "It's almost like you like mistakes more than the correct answers." Though the student said this in jest, it might be the best compliment I have received as a teacher. If my students think I like mistakes more than correct answers, then I have made my classroom a safe space where mistakes are celebrated, and that's such a win for me. Although I am now one of the authors of a book about celebrating mathematical mistakes, learning how to value mistake making and integrate mistakes into my classroom is something I will work on for my lifetime.

A REFLECTION FROM NATASHA

I have a distinct memory from teaching fifth-grade mathematics. I can see this girl from the class who was often very shy and sometimes did not enjoy speaking. We had been working on a rich mathematical task together (although I've long since forgotten what it was even about), and the girl started giggling, and a huge smile spread across her face. She appeared antsy, like she could not sit still. I asked her, not necessarily in front of the entire class, but a bit more loudly than a one-on-one conversation would require, "What's going on? You seem like you want to share something!" She replied, "I do! I made the *best* mistake!" for the class to hear. We proceeded to unpack her mistake as a whole class, and even though I do not remember what her mistake was, that does not matter for this story. What I will never forget is her joy and delight when she finally felt comfortable enough to (1) share out loud in the class, (2) share out loud that she had made a mistake, and (3) deem her mistake the best mistake (relative to others we had been investigating). At the end of the year, the girl made me something in her art class out of wooden craft sticks and yarn. When she gave it to me, it was obvious she had made a few adjustments to her work along the way. She told me, "I made this for you so you could see that I like making mistakes in all the things I do." It's one of my most treasured mementos from teaching to this day.

★ ★ ★

Revisiting the Big Ideas From Each Chapter

Over the course of this book, we have all journeyed together toward valuing and celebrating mistakes in the mathematics classroom. We hope to continue to grow with you as educators who expect, inspect, and respect these mistakes (Seeley, 2016). The pursuit of growing in celebrating and using mistakes in the classroom is a lifelong journey. And we hope this book has helped you, wherever you are in that journey, whether you are just starting to support mistakes in the classroom or are fine-tuning your craft of supporting and revising mistakes.

In our final chapter together, we want to bring everything full circle by revisiting the big ideas we set forth in each chapter, which are listed in table 10.1. We hope that while reading this book, you have made micro-adjustments—or perhaps major ones—to your practice to move toward asset-based views of mistakes. If so, pat yourself on the back! Micro-adjustments over time lead to major changes in your practice and student outcomes. If you have loftier goals, perhaps you have fully embraced several of the big ideas laid out in table 10.1 and are wondering what the classroom would look like if you were to adopt *all* the big ideas.

Table 10.1: Big Ideas From Each Chapter

Chapter	Big Idea
1	Developing mathematicians means helping students shift from deficit-based views to asset-based views of mathematical mistakes.
2	Defining beauty and power in mathematics requires teachers to think about how mathematical mistakes align with beauty and power. Seeing mistakes as beautiful and powerful mathematics provides opportunities for thinking about how to use mistakes in the classroom in different ways.
3	Types of mistakes include factual, procedural, and conceptual. Being able to identify which type your student made, considering where they are on their learning trajectory and what the lesson goal is, helps you determine how to address the mistake.
4	We are all mathematicians. All mathematical mistakes that help people learn are worthy of celebration. This chapter is celebratory of different mathematical mistakes. This explicit celebration is a tool for supporting the development of your and your students' identities as mathematicians.

continued →

5	A foundation for investigating mathematical mistakes in your classroom can begin with using "unknown student work" that contains a mistake and carefully selected tasks.
6	The Changing Minds Task Structure is a useful way to support the use of both incorrect and correct solutions in the same task.
7	The This or That Task Structure can be used to support the use of mistakes as a way of deepening conceptual understanding.
8	Supporting invented mathematical notation and language is a way to support creativity that will naturally result in mistake making. Yet some of the mistakes that happen with supporting invented notation and language in the mathematics classroom are not really mistakes at all.
9	Using authentic play in mathematics goes beyond just using mathematical games. Mathematical games are not necessarily real play. Rather, intentionally integrating playful experiences (for example, games with tenets of play, and intellectual play) at all grade bands helps create a space where students feel safe to make mistakes to learn from.
10	There are a variety of organic and planned ways you can use the ideas from this book in a mathematics class to support mistake making in asset-based ways.

Next, we present a vignette describing a classroom that has fully adopted all the big ideas. We pause throughout the vignette to highlight the various big ideas. We reflect on things the teacher does that you can easily incorporate into your classroom as well.

We present Mrs. W, a fifth-grade teacher whose class is working on decimal multiplication. Specifically, the students have finished their work on whole numbers times a decimal (and vice versa) and are now working on multiplication with two decimal numbers (hundredths and tenths). Her classroom is set up so students' desks are grouped in pairs or trios and every student has a partner or partners with whom to work. The configuration communicates that students are expected to work together, and this orientation of her room only changes on assessment days.

Mrs. W structures her lessons similarly each day: (1) warm-up, (2) main activity, (3) synthesis, and (4) exit ticket. Typically, students then leave mathematics class for specials, and when they return, if time permits, she has center time where students can play certain games to focus on skill fluency. We take a peek into Mrs. W's class at the beginning of mathematics class, and we see that she has the warm-up on the board: *What is ½ of 1.50?*

Mistakes in Action

Students are expected to work on warm-ups mentally if they are able, but they always have the option to use a whiteboard for scratch work or use manipulatives if that supports their thinking. As the students work on this warm-up in their groups, Mrs. W sets their timer for three minutes. She notices that some students are using base 10 blocks, some have drawn pictures of base 10 blocks on their whiteboards, and some are not using any tools (likely using mental mathematics). She sees a variety of approaches to solving the warm-up in terms of tools and strategies used, but it seems that most students are getting the same answer. As the timer buzzes, Mrs. W pulls the class together.

Mrs. W: Thank you all for working so diligently on this warm-up. I know that many of you had the same answer, but I saw at least three different ways that you all got the answer. First, let's quickly confirm the solution because that's not really what I'm going to focus on since you all know I love a good strategy more than I love a correct solution. Cera, what answer did your team get?

Cera: We got 75 hundredths.

Mrs. W: What other answers did we get? (No other student shares.) *OK, let's get to the really cool part of mathematics—the strategies! Ami, how did you and your team solve this problem?*

Ami: We used our [base 10] blocks. We used one of the hundreds blocks and then five of the rods. Because that would be 1 and 50, which is 1.50.

Mrs. W: OK, so you used manipulatives to solve it. Why did you decide to use the hundreds block to represent the 1?

Ami: Well, maybe that's wrong. We thought that since we had three numbers—1, 5, and 0—in our number that we needed three kinds of blocks—the hundreds, the tens rods, and the ones cubes. So, we just let the number 1 be the hundreds, the 5 be our rods, and the 0 be our ones.

Mrs. W: Then what did you do?

Ami: Then we took each of those blocks and split them into two piles because it said to take one-half of that number. So, half of the hundreds is a 50, or five rods. And half of the five rods is two rods and five small cubes, so 25. So, then we have five rods, two rods, and five small cubes, or 75.

Mrs. W: Interesting! Is that 75 or is that 75 hundredths? Think about that! Lonny, your team got the same answer but differently. How did your team think about the problem?

celebrating mathematical mistakes

Lonny: We got the same answer—75 cents. But we thought about money because, like, 1.50 looks like a dollar fifty. And half of a dollar is 50 cents. Half of 50 cents is a quarter. And then 50 cents and a quarter, that is 75 cents.

Mrs. W: So, your team did that all in your heads?

Lonny: We each kinda thought about it a second. Then we talked and thought that it made sense, so we did it that way in our heads.

Mrs. W: I love it. Finally, I want to hear from Raz's group. But we've only got a few more minutes in warm-up, so make sure we cover our strategy quickly.

Raz: We made a mistake, but we figured it out and still got 0.75. We did long division because we know that one-half of 1.50 is the same thing as 1.50 divided by 2. So, we did 1.50 divided by 2, and we made a mistake when we divided, but then Liz helped me, and we got it right.

Liz: Yeah, I showed Raz how he did 2 times 7 is 14, but then when he went to write it, he forgot about his decimal point. So, he thought the answer was 75 and not 0.75, and that can't be the answer because 75 is way too high.

Mrs. W: Thank you for explaining that. I just want to point out that Raz's team was able to use the mistake to reason through their answer that 75 couldn't possibly be the answer because that seemed too high, so they went back, examined their mistake, corrected it, and figured out the right answer. I love how your mistake helped you get to the right answer! That is so cool that something that was a mistake turned into helping you solve the problem! Thank you, Raz's team. Great work! Now, let's transition over to our lesson for today.

Notice how Mrs. W acknowledged the mistake Raz's team made and positioned it as an asset to learning. By doing so, she embraced the big idea from chapter 1 (page 12) of shifting from a deficit-based perspective to an asset-based perspective when dealing with mathematical mistakes. Additionally, this was a great way to celebrate a mistake and highlight its power in learning, which is the big idea from chapter 2 (page 21). When Mrs. W took an asset-based approach that highlighted how a mistake can help learning, she also positioned Ami as a young mathematician. In fact, facilitating all students as young mathematicians who can create strategies and make mistakes supports the big ideas of chapter 1 and chapter 4 (mathematical mistakes that help people learn are worthy of celebration; page 63).

Mistakes in Action

Mrs. W projects the following task on the board and asks a student to read it out loud: *Sasha is a chemist. In her current experiment, she needs a quarter of 0.2 liter of white vinegar. Write a mathematical equation to represent this problem, and determine how much vinegar Sasha needs for her experiment.*

> Mrs. W: What questions do you all have about this problem?
>
> Rainey: What's a chemist?
>
> Mrs. W: That is a scientist who studies chemistry. What other questions do you have?
>
> Zoe: Can we do an experiment?
>
> Mrs. W: Let me rephrase. What mathematical questions do you have about this problem? (No other students ask a question.) You all know the routine; let's work in our duos or trios. You have manipulatives if you need them. I'll give you seven minutes on the timer to get started, and I will stop you to share strategies, even unfinished ones, at some point—not answers! Answers will be for later. Get to work.

Students start talking to one another as Mrs. W circulates through the room, noting what strategies students are choosing to solve the problem. After about four minutes, Mrs. W notices that many students are approaching the problem with mistakes in their work. She notices that these mathematical mistakes are conceptual. She decides to pause the groups and have some share their strategies to see if that can help them adjust any of their reasoning toward a correct strategy. Before pausing, Mrs. W selects two groups to share their work because their mistakes are ubiquitous to the class and potentially accessible to other students. She has decided, depending on how the conversation progresses, she might show a third mistake based on what her students have done in the past that can move them toward a correct strategy.

Notice that Mrs. W has done some examination and categorization of mistakes. She has recognized that the mistakes she is seeing are ubiquitous and worthy of investigation. She identifies the mistakes as conceptual, thus embracing the big idea from chapter 3 (page 39). Centering a mistake that addresses conceptual, rather than procedural or factual, understanding provides opportunities for deeper discussion in the mathematics classroom.

> Mrs. W: I know we still have a few minutes on the timer. But I want us to pause and talk about our strategies. Siera, will your group share what strategy you used to approach this problem?
>
> Siera: Yeah, so we thought about our warm-up. We thought the [base 10] blocks would help us, so we got two rods to show the point two. And then we are going to repeat that 25 times because 25 is a quarter.
>
> Mrs. W: I saw at least one other group do the same thing. Wilder, your group did something similar, right?
>
> Wilder: Yup. A quarter means 25, so 25 times point two. That's what we did, but we drew pictures of the blocks. We didn't actually pull them out.
>
> Mrs. W: What do the other groups think about this approach? What questions do you have for these two groups about their strategy?

Mrs. W hopes a student or group of students will acknowledge that a quarter represents ¼ instead of 25. If one does, she will move on to another mistake she saw during her circulation. If not, she plans to pull out unknown student work for a changing minds task.

Mrs. W decides to use the big ideas from chapters 5 and 6 (pages 83 and 103) together. Although she chooses to use the changing minds task, she could also implement a this or that task, which is the big idea from chapter 7 (page 123). Many of the ideas we have presented can be seamlessly merged into one pedagogical move.

Next, Mrs. W engages with Raz on multiplying with decimals in the context of money.

> Raz: I think what they did makes sense because a quarter is 25 cents, so I like it.
>
> Mrs. W: I really love how all of you are thinking. And I see some similarities in your strategies. Thank you all for sharing these ideas! I noticed that what Siera's and Wilder's groups did was very similar to what one of my students did last year, but this student of mine ended up changing her mind. I want to show you what she did.

Mrs. W writes on the board what appears in figure 10.1 (page 183).

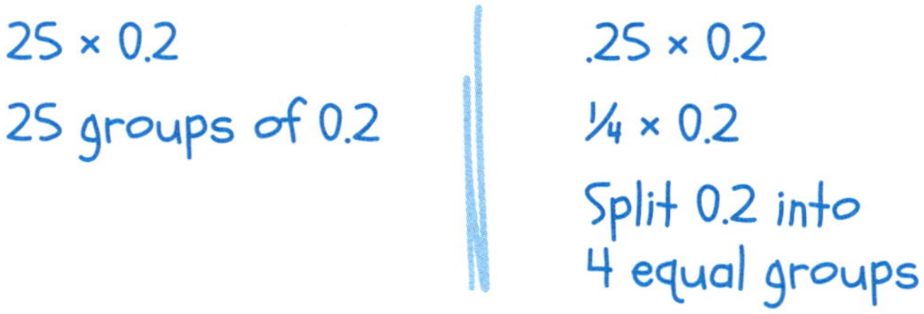

Mistakes in Action

$$25 \times 0.2$$
25 groups of 0.2

$$.25 \times 0.2$$
$$\tfrac{1}{4} \times 0.2$$
Split 0.2 into 4 equal groups

Figure 10.1: Ace's changing her mind on how to solve one-quarter of 2 tenths.

Mrs. W: Let's call this student Ace. On the left is what Ace did first. Then, Ace decided to change her mind, and what you see on the right is what she did second. Which way do you think is correct? Why? Talk to your teammates.

Cera: Our team sees what she did. Her work on the left is like what Wilder and Siera did, but then she changed her mind to the work on the right, and we think that is correct.

Raz: Yeah, we think the work on the right is correct because, even though a quarter is 25 cents, like, I get why what is on the left makes sense. But 25 cents is really a quarter of a dollar, which is like taking a hundred pennies and splitting them into four groups.

Wilder: I see. So a quarter is the same thing as point two five, right?

Mrs. W: Twenty-five hundredths is the same thing as a quarter, yes. What questions do you have about Ace's work?

Raz: Yeah, why did she change her mind?

Mrs. W: Good question. Ace was working with her teammates, and one of the teammates said they didn't think a quarter was the same thing as 25. So, they talked about what a quarter would be if it wasn't 25, and they came up with the work on the right.

Raz: OK.

Mrs. W: Now that we've seen Ace's work, do we feel like we have some more information about the problem and can move forward?

The class agrees and gets back to work. Mrs. W sets the timer for five more minutes. As students are working, Mrs. W circulates through the room and selects a few groups she wants to share with the class. She chooses groups that accurately capture the majority of the thinking in the room but also have different representations and equations or have some mistakes worth discussing. She asks these groups to put their work on vertical whiteboards throughout the room. After five minutes, Mrs. W has the selected groups stand by their vertical whiteboards. Groups not selected rotate to each whiteboard and ask questions about the work presented. Then, Mrs. W asks a group that was not selected to share a vertical presentation with the whole class as if it was their own—thus encouraging everyone to be invested in the presentation and work. We pick back up after this, when Mrs. W is revoicing and synthesizing the work presented.

> Mrs. W: So, what I heard many of you saying is that we all agreed the way we wanted to approach this was 25 hundredths times 2 tenths. But where I heard some discrepancy is in what our answer might be. Some groups think the answer is 5 tenths. Some groups said 5, another group thinks the answer is (looking at their work to capture the correct student wording) *point zero 5*, and one group had this same answer, but they called it 5 hundreds. So, we have agreed on the same problem to solve, but we have different answers.
>
> Unfortunately, we do not have time to determine who is correct before we leave for specials, so I want your answer on an exit ticket. (All students grab a sticky note while Mrs. W writes the answers that have been presented on the board.) *On your exit ticket, write the answer you think is correct and your name. We will start our warm-up tomorrow by addressing which is correct by modeling our answer with graph paper, which is something we haven't seen yet, but some of my students last year found it really helpful!*

Mrs. W does not correct students when they say "5 hundreds" instead of "5 hundredths," nor does she push students to say "5 tenths" instead of "point zero 5" *yet*. Yes, being precise in mathematics is important, but Mrs. W is striking a delicate balance in knowing not to correct a factual mistake right away and allowing students to think and unpack a conceptual mistake. Although saying "5 hundreds" is a mistake, it could very well take care of itself when the solution is presented the next day. The second statement is something that some teachers identify as a mistake: "It is not correct to say 'point zero 5.'" But in daily life, people use this regularly, so it is not something Mrs. W has prioritized. Because students will likely experience this language outside the classroom, they should know what it means and how to use it. Here, Mrs. W has fully embraced the big idea from chapter 8 (page 135), recognizing that not all mistakes are actually mistakes and supporting students' invented or constructed language as she supports them toward academic language.

When the students return to class, they complete mathematics fluency games that develop procedural skills. Mrs. W has selected games as a playful way to revisit materials that can build factual and procedural knowledge for students. She hopes students will learn from their mistakes as they play the games. Although we do not see Mrs. W implement this fluency time, we can see from the preceding excerpts that students are prepared for that opportunity, and thus, she has implemented the big idea from chapter 9 (page 155), where students are supported in play in their classroom as a space for making and using mistakes.

Concluding Remarks

It is apparent from this vignette that Mrs. W spent considerable time and effort developing community norms and expectations in her classroom at the beginning of the year to make many of these moves possible. It is also obvious that there is still room to grow. Like all teachers, Mrs. W is not perfect. Therefore, to have a truly inclusive classroom where mistakes are valued, expected, and inspected, educators must expect to make their own mistakes.

We hope you, like Mrs. W, are brave in your journey toward creating a classroom where mistakes are celebrated and lead to deeper mathematical understanding. Happy mistake making! May your classroom be a celebration of mistakes that lead to beautiful and powerful mathematics.

Reflection Questions

Use the following questions for reflecting on the ideas in chapter 10.

1. Write a vignette or share a story from your mathematics classroom that illustrates asset-based views of mathematical mistakes.

2. What is your favorite big idea from this book? Why is it your favorite, and how will you use it in your classroom?

3. What is one pedagogical tool from this book that you want to try in your mathematics classroom (for example, unknown student work, this or that tasks, or changing minds tasks)? How will you use it?

4. Why do you think celebrating mathematical mistakes is important? How will you celebrate mathematical mistakes in the future?

Chapter 10 Application Guide

In this chapter, we discussed how all this book's big ideas about mathematical mistakes can be used in organic and planned ways in the classroom. Use the following application guide to connect the chapter's themes to your classroom.

Chapter Theme	Connection and Application to Your Practice
Mathematical mistakes can be used and celebrated in the classroom in a variety of ways.	Although we have shared ten big ideas about using and celebrating mathematical mistakes in your classroom in asset-based ways, it is unrealistic that you will use all the big ideas at once. Instead, pick one or two big ideas about mathematical mistakes to focus on, and you will notice that you inherently start to use other big ideas.
Celebrate and use mathematical mistakes.	Pick the idea about mathematical mistakes from part 1 that resonates the most with you. Write it on a card and put it somewhere you will see it every day (for example, in your planner, next to your computer) as a reminder to yourself. Then pick a task structure or pedagogical idea from part 2 and try something new in your classroom!

Epilogue

Thomas Edison said, "I have gotten a lot of results! I know several thousand things that won't work" (Ratcliffe, 2016). We hope that your students find "several thousand things that won't work" as they learn about mathematics, and that you celebrate them as mathematicians for that work. For us, this book has laid the foundation for transforming our perspectives on mistake making in our mathematics classrooms. However, the journey toward this transformation will likely take you longer than it took to read this book. For each of us, the journey has been years in the making and still continues. We both started our teaching journeys with very different-looking mathematics classrooms than the ones we cultivate today.

In some ways, celebrating mathematical mistakes extends beyond mathematics alone. Consider the ways in which you make mistakes on this journey toward changing your practice; the fact that you make them at all is a sign that you are courageous in taking up new approaches to teaching. Further, it is a sign that you are on the verge of an amazing transition in your practice—one where students will begin to joyfully look for ways to use their mistakes to propel their learning (something all the great mathematicians and scientists are adept at, as shown in Thomas Edison's quote).

As we leave you, we want to charge you with one final task, and that is to allow yourself and your students grace. The journey toward celebrating mathematical mistakes will be long, and perhaps arduous at times, but the payoff at the end is worth it.

References and Resources

Adiredja, A. P., & Louie, N. (2020). Untangling the web of deficit discourses in mathematics education. *For the Learning of Mathematics*, *40*(1), 42–46.

Aguirre, J., Mayfield-Ingram, K., & Martin, D. B. (2013). *The impact of identity in K–8 mathematics learning and teaching: Rethinking equity-based practices*. Reston, VA: National Council of Teachers of Mathematics.

Allen, K., & Schnell, K. (2016). Developing mathematics identity. *Mathematics Teaching in the Middle School*, *21*(7), 398–405.

Aristotle. (c. 350 BC). *Metaphysics* (W. D. Ross, Trans.). Accessed at www.documentacatholicaomnia.eu/03d/-384_-322,_Aristoteles,_13_Metaphysics,_EN.pdf on February 21, 2024.

Baker, K., Jessup, N. A., Jacobs, V. R., Empson, S. B., & Case, J. (2020). Productive struggle in action. *Mathematics Teacher: Learning and Teaching PK–12*, *113*(5), 361–367.

Barlow, A. T., Gerstenschlager, N. E., & Harmon, S. E. (2016). Learning from the unknown student. *Teaching Children Mathematics*, *23*(2), 74–81.

Barlow, A. T., Watson, L. A., Tessema, A. A., Lischka, A. E., & Strayer, J. F. (2018). Inspection-worthy mistakes: Which? And why? *Teaching Children Mathematics*, *24*(6), 384–391.

Battey, D., & Leyva, L. A. (2016). A framework for understanding whiteness in mathematics education. *Journal of Urban Mathematics Education*, *9*(2), 49–80.

Bay-Williams, J., & Kling, G. (2019). *Math fact fluency: 60+ games and assessment tools to support learning and retention*. Arlington, VA: ASCD.

Bieda, K. N., & Staples, M. (2020). Justification as an equity practice. *Mathematics Teacher: Learning and Teaching PK–12*, *113*(2), 102–108.

Bishop, J. P., Lamb, L. L., Philipp, R. A., Schappelle, B. P., & Whitacre, I. (2011). First graders outwit a famous mathematician. *Teaching Children Mathematics*, *17*(6), 350–358.

Bishop, J. P., Lamb, L. L., Philipp, R. A., Whitacre, I., & Schappelle, B. P. (2016a). Leveraging structure: Logical necessity in the context of integer arithmetic. *Mathematical Thinking and Learning*, *18*(3), 209–232.

Bishop, J. P., Lamb, L. L., Philipp, R. A., Whitacre, I., & Schappelle, B. P. (2016b). Unlocking the structure of positive and negative numbers. *Mathematics Teaching in the Middle School*, *22*(2), 84–91.

Bishop, J. P., Lamb, L. L., Philipp, R. A., Whitacre, I., & Schappelle, B. P. (2018). Students' thinking about integer open number sentences. In L. Bofferding & N. M. Wessman-Enzinger (Eds.), *Exploring the integer addition and subtraction landscape: Perspectives on integer thinking* (pp. 47–71). Cham, Switzerland: Springer.

Bishop, J. P., Lamb, L. L., Philipp, R. A., Whitacre, I., Schappelle, B. P., & Lewis, M. L. (2014). Obstacles and affordances for integer reasoning: An analysis of children's thinking and the history of mathematics. *Journal for Research in Mathematics Education*, *45*(1), 19–61.

Boaler, J. (2016). *Mathematical mindsets: Unleashing students' potential through creative math, inspiring messages, and innovative teaching.* San Francisco: Jossey-Bass.

Bofferding, L. (2010). Addition and subtraction with negatives: Acknowledging the multiple meanings of the minus sign. In P. Brosnan, D. Erchick, & L. Flevares (Eds.), *Proceedings of the 32nd annual meeting of the North American chapter of the International Group for the Psychology of Mathematics Education* (pp. 703–710). Columbus, OH: The Ohio State University.

Bofferding, L. (2012). *–5 – –5 is like 5 – 5: Analogical reasoning with integers* [Paper presentation]. American Educational Research Association, Vancouver, British Columbia, Canada.

Bofferding, L. (2014). Negative integer understanding: Characterizing first graders' mental models. *Journal for Research in Mathematics Education*, *45*(2), 194–245.

Bofferding, L., Aqazade, M., & Farmer, S. (2018). Playing with integer concepts: A quest for structure. In L. Bofferding & N. M. Wessman-Enzinger (Eds.), *Exploring the integer addition and subtraction landscape: Perspectives on integer thinking* (pp. 3–23). Cham, Switzerland: Springer.

Bofferding, L., & Hoffman, A. (2015). Comparing negative integers: Issues of language. In K. Beswick, T. Muir, & J. Wells (Eds.), *Proceedings of the 39th Psychology of Mathematics Education Conference* (Vol. 1, p. 150). Hobart, Australia: PME.

Bofferding, L., & Wessman-Enzinger, N. (2017). Subtraction involving negative numbers: Connecting to whole number reasoning. *The Mathematics Enthusiast*, *14*(1–3), 241–262.

Boone, D. J. (2007). *A Picasso or a preschooler? Ways of seeing the "child as artist"* [Conference presentation]. Philosophy of Education Society of Australasia, Wellington, New Zealand. Accessed at https://eprints.qut.edu.au/12477/1/DBoone_2007_PESA_conference_paper.pdf on May 29, 2024.

Britt, B. A. (2014). *Mastering basic math skills: Games for kindergarten through second grade.* Reston, VA: National Council of Teachers of Mathematics.

Britt, B. A. (2015). *Mastering basic math skills: Games for third through fifth grade.* Reston, VA: National Council of Teachers of Mathematics.

Brown, R. (2017). Using collective argumentation to engage students in a primary mathematics classroom. *Mathematics Education Research Journal*, *29*, 183–199.

Buckley, S., Reid, K., Lipp, O. V., Goos, M., Bethune, N., & Thomson, S. (2021). Addressing mathematics anxiety in primary teaching. In A. Carroll, R. Cunnington, & A. Nugent (Eds.), *Learning under the lens: Applying findings from the science of learning to the classroom* (pp. 78–92). New York: Routledge.

Burghardt, G. M. (2011). Defining and recognizing play. In A. D. Pellegrini (Ed.), *The Oxford handbook of the development of play* (pp. 9–18). New York: Oxford University Press.

Carpenter, C. H., & Wessman-Enzinger, N. M. (2018). Grade 5 students' negative integer multiplication strategies. In T. E. Hodges, G. J. Roy, & A. M. Tyminski (Eds.), *Proceedings of the 40th annual meeting of the North American chapter of the International Group for the Psychology of Mathematics Education* (pp. 139–146). Greenville, SC: University of South Carolina & Clemson University.

Carpenter, T. P., Fennema, E., Franke, M. L., Levi, L., & Empson, S. B. (2015). *Children's mathematics: Cognitively guided instruction* (2nd ed.). Portsmouth, NH: Heinemann.

Celedón-Pattichis, S., Borden, L. L., Pape, S. J., Clements, D. H., Peters, S. A., Males, J. R., et al. (2018). Asset-based approaches to equitable mathematics education research and practice. *Journal for Research in Mathematics Education, 49*(4), 373–389.

Dacey, L., Gartland, K., & Lynch, J. B. (2015a). *Well played: Building mathematical thinking through number games and puzzles, grades K–2*. New York: Routledge.

Dacey, L., Gartland, K., & Lynch, J. B. (2015b). *Well played: Building mathematical thinking through number games and puzzles, grades 3–5*. New York: Routledge.

Dacey, L., Gartland, K., & Lynch, J. B. (2024). *Well played: Building mathematical thinking through number and algebraic games and puzzles, grades 6–8*. New York: Routledge.

Day, E. (Host). (2018–present). *How to fail with Elizabeth Day* [Audio podcast]. Failing With Friends.

Di Martino, P., & Zan, R. (2013). Where does fear of maths come from? Beyond the purely emotional. In B. Ubuz, C. Haser, & M. Mariotti (Eds.), *Proceedings of the eighth congress of the European Society for Research in Mathematics Education* (pp. 1272–1450). Ankara, Turkey: Middle East Technical University.

Dweck, C. S. (2006). *Mindset: The new psychology of success*. New York: Random House.

Dweck, C. S., & Yeager, D. S. (2019). Mindset: A view from two eras. *Perspectives on Psychological Science, 14*(3), 481–496.

Ellis, J., Fosdick, B. K., & Rasmussen, C. (2016). Women 1.5 times more likely to leave STEM pipeline after calculus compared to men: Lack of mathematical confidence a potential culprit. *PLoS ONE, 11*(7), Article e0157447.

English, L. D. (1998). Reasoning by analogy in solving comparison problems. *Mathematical Cognition, 4*(2), 125–146.

English, L. D. (2004). Mathematical and analogical reasoning in early childhood. In L. D. English (Ed.), *Mathematical and analogical reasoning of young learners* (pp. 1–22). Mahwah, NJ: Erlbaum.

Featherstone, H. (2000). "-Pat + Pat = 0": Intellectual play in elementary mathematics. *For the Learning of Mathematics, 20*(2), 14–23.

Featherstone, H., Crespo, S., Jilk, L. M., Oslund, J. A., Parks, A. N., & Wood, M. B. (2011). *Smarter together! Collaboration and equity in the elementary math classroom*. Reston, VA: National Council of Teachers of Mathematics.

Fosnot, C. T., & Dolk, M. (2001). *Young mathematicians at work: Constructing number sense, addition, and subtraction*. Portsmouth, NH: Heinemann.

Ganley, C. M., Schoen, R. C., LaVenia, M., & Tazaz, A. M. (2019). The construct validation of the math anxiety scale for teachers. *AERA Open, 5*(1), 1–16.

GermanPod101. (2021, June 10). *30 German proverbs and idioms to speak like a native* [Blog post]. Accessed at www.germanpod101.com/blog/2021/06/10/best-german-proverbs on February 21, 2024.

Ginsburg, H. P. (2006). Mathematical play and playful mathematics: A guide for early education. In D. G. Singer, R. M. Golinkoff, & K. Hirsh-Pasek (Eds.), *Play = learning: How play motivates and enhances children's cognitive and social-emotional growth* (pp. 145–165). New York: Oxford University Press. https://doi.org/10.1093/acprof:oso/9780195304381.003.0008

Hammond, Z. (2015). *Culturally responsive teaching and the brain: Promoting authentic engagement and rigor among culturally and linguistically diverse students*. Thousand Oaks, CA: Corwin.

Herbel-Eisenmann, B. A., & Breyfogle, M. L. (2005). Questioning our patterns of questioning. *Mathematics Teaching in the Middle School, 10*(9), 484–489.

Herbel-Eisenmann, B. A., Steele, M. D., & Cirillo, M. (2013). (Developing) teacher discourse moves: A framework for professional development. *Mathematics Teacher Educator, 1*(2), 181–196.

Hertel, J. T., & Wessman-Enzinger, N. M. (2017). Examining Pinterest as a curriculum resource for negative integers: An initial investigation. *Education Sciences, 7*(2), 45. https://doi.org/10.3390/educsci7020045

Jansen, A. (2020). Rough-draft thinking and revising in mathematics. *Mathematics Teacher: Learning and Teaching PK–12, 113*(12), 107–110.

Kaplan, R. (2000). *The nothing that is: A natural history of zero*. New York: Oxford University Press.

Karp, K. S., Bush, S. B., & Dougherty, B. J. (2014). 13 rules that expire. *Teaching Children Mathematics, 21*(1), 18–25.

Kilani, I. (2023). Fermat's Last Theorem: A historical and mathematical overview. *ESS Open Archive*. https://doi.org/10.22541/essoar.169447369.92321625/v1

Kilpatrick, J., Swafford, J., & Findell, B. (Eds.). (2001). *Adding it up: Helping children learn mathematics*. Washington, DC: National Academies Press. https://doi.org/10.17226/9822

Kshetree, M. P. (2018). Nature of mathematical content as a contributing factor for students' mathematical errors. *International Journal of Mathematical Trends and Technology, 58*(4), 309–317.

Kurth, L. A., Anderson, C. W., & Palincsar, A. S. (2002). The case of Carla: Dilemmas of helping *all* students to understand science. *Science Education, 86*(3), 287–313.

Lamb, L. L., Bishop, J. P., Philipp, R. A., Whitacre, I., & Schappelle, B. P. (2018). A cross-sectional investigation of students' reasoning about integer addition and subtraction: Ways of reasoning, problem types, and flexibility. *Journal for Research in Mathematics Education, 49*(5), 575–613.

Leighton, J. P., Guo, Q., & Tang, W. (2021). Measuring preservice teachers' attitudes towards mistakes in learning environments. *Learning Environments Research, 25*, 287–304.

Liljedahl, P. (2015, February). *Emotions as an orienting experience* [Paper presentation]. Ninth congress of the European Society for Research in Mathematics Education, Prague, Czech Republic.

Lischka, A. E., Gerstenschlager, N. E., Stephens, D. C., Strayer, J. F., & Barlow, A. T. (2018). Making room for inspecting mistakes. *Mathematics Teacher, 111*(6), 432–439.

Louie, N., Adiredja, A. P., & Jessup, N. (2021). Teacher noticing from a sociopolitical perspective: The FAIR framework for anti-deficit noticing. *ZDM: Mathematics Education, 53*(1), 95–107.

Marche, S. (2023). *On writing and failure, or, On the peculiar perseverance required to endure the life of a writer*. Windsor, Ontario, Canada: Biblioasis.

Martínez, A. A. (2006). *Negative math: How mathematical rules can be positively bent*. Princeton, NJ: Princeton University Press.

Mendelowitz, D. M. (1963). *Children are artists: An introduction to children's art for teachers and parents* (2nd ed.). Stanford, CA: Stanford University Press.

Miner, R. N. (2013). *Fermat's Last Theorem* [Unpublished thesis]. University of Redlands. https://core.ac.uk/download/pdf/217142344.pdf

Mis-. (n.d.). In *Merriam-Webster's online dictionary*. Accessed at www.merriam-webster.com/dictionary/mis- on March 29, 2024.

Mozzochi, C. J. (2004). Essay review: Quest and conquest—Proof of Fermat's Last Theorem. *Annals of Science, 61*(1), 119–126.

Muhammad, G. (2020). *Cultivating genius: An equity framework for culturally and historically responsive literacy*. New York: Scholastic.

Muzundu, K. (2024). A result in odd powers in Fermat's Last Theorem. *Mathematics Letters, 10*(1), 1–6.

National Council of Teachers of Mathematics. (2014). *Principles to actions: Ensuring mathematical success for all.* Reston, VA: Author.

National Governors Association Center for Best Practices & Council of Chief State School Officers. (2010). *Common Core State Standards for mathematics.* Washington, DC: Authors. Accessed at https://corestandards.org/wp-content/uploads/2023/09/Math_Standards1.pdf on February 21, 2024.

Nieder, A. (2016). Representing something out of nothing: The dawning of zero. *Trends in Cognitive Sciences, 20*(11), 830–842.

O'Dell, J. R. (2018). The interplay of frustration and joy: Elementary students' productive struggle when engaged in unsolved problems. In T. E. Hodges, G. J. Roy, & A. M. Tyminski (Eds.), *Proceedings of the 40th annual meeting of the North American chapter of the International Group for the Psychology of Mathematics Education* (pp. 938–945). Greenville, SC: University of South Carolina & Clemson University.

Olson, S., & Riordan, D. G. (2012, February). *Report to the president: Engage to excel—Producing one million additional college graduates with degrees in science, technology, engineering, and mathematics.* Washington, DC: Executive Office of the President.

Orlin, B. (2022). *Math games with bad drawings: 75 ¼ simple, challenging, go-anywhere games—and why they matter.* New York: Black Dog & Leventhal.

Parks, A. N. (2015). *Exploring mathematics through play in the early childhood classroom.* New York: Teachers College Press.

Parks, A. N. (2020). Creating joy in PK–grade 2 mathematics classrooms. *Mathematics Teacher: Learning and Teaching PK–12, 113*(1), 61–64.

Parrish, S. (2010). *Number talks: Helping children build mental math and computation strategies, grades K–5.* Sausalito, CA: Math Solutions.

Parrish, S. D. (2011). Number talks build numerical reasoning. *Teaching Children Mathematics, 18*(3), 198–206.

Peterson, J. C. (1972). Fourteen different strategies for multiplication of integers or why (–1) (–1) = +1. *The Arithmetic Teacher, 19*(5), 396–403.

Petit, M. M., Laird, R. E., Marsden, E. L., & Ebby, C. B. (2016). *A focus on fractions: Bringing research to the classroom* (2nd ed.). New York: Routledge.

Poincaré, H. (1910). Mathematical creation. *The Monist, 20*(3), 321–335.

Ratcliffe, S. (Ed.). (2016). *Oxford essential quotations* (4th ed.). Oxford, England: Oxford University Press. Accessed at www.oxfordreference.com/display/10.1093/acref/9780191826719.001.0001/q-oro-ed4-00003960# on March 28, 2024.

Ravitch, D. (2010). *The death and life of the great American school system: How testing and choice are undermining education.* New York: Basic Books.

Rumsey, C., & Langrall, C. W. (2016). Promoting mathematical argumentation. *Teaching Children Mathematics, 22*(7), 412–419.

Sagal, P. (Host). (2015, October 24). Not my job: We quiz cosmos expert Neil deGrasse Tyson on cosmetology [Radio show segment]. In *Wait wait . . . don't tell me!* NPR. Accessed at www.npr.org/2015/10/24/450994221/not-my-job-we-quiz-cosmos-expert-neil-degrasse-tyson-on-cosmetology on February 28, 2024.

Seeley, C. L. (2016). *Making sense of math: How to help every student Become a mathematical thinker and problem solver.* Arlington, VA: ASCD.

Sheffield, L. J., Meissner, H., & Foong, P. Y. (2004, July). *Developing mathematical creativity in young children* [Paper presentation]. Tenth International Congress on Mathematical Education, Copenhagen, Denmark.

Sinclair, N. (2004). The roles of the aesthetic in mathematical inquiry. *Mathematical Thinking and Learning, 6*(3), 261–284.

Singh, S. (2002). *Fermat's Last Theorem.* New York: HarperCollins.

Smith, M. S., & Stein, M. K. (1998). Reflections on practice: Selecting and creating mathematical tasks—From research to practice. *Mathematics Teaching in the Middle School, 3*(5), 344–350.

Song, E., & Id-Deen, L. (2023). *Disrupting injustice: Navigating critical moments in the classroom*. Reston, VA: National Council of Teachers of Mathematics.

Stoehr, K. J., & Olson, A. M. (2023). Elementary prospective teachers' visions of moving beyond mathematical anxiety. *Mathematics Education Research Journal, 35*(1), 133–152.

Su, F. (2020). *Mathematics for human flourishing*. New Haven, CT: Yale University Press.

Sun, K. L. (2018). The role of mathematics teaching in fostering student growth mindset. *Journal for Research in Mathematics Education, 49*(3), 330–355.

Townsend, C., Slavit, D., & McDuffie, A. R. (2018). Supporting all learners in productive struggle. *Mathematics Teaching in the Middle School, 23*(4), 216–224.

Ukpokodu, O. N. (2011). How do I teach mathematics in a culturally responsive way? Identifying empowering teaching practices. *Multicultural Education, 19*(3), 47–56.

Vaartio-Rajalin, H., Santamäki-Fischer, R., Jokisalo, P., & Fagerström, L. (2021). Art making and expressive art therapy in adult health and nursing care: A scoping review. *International Journal of Nursing Sciences, 8*(1), 102–119.

Wagganer, E. L. (2015). Creating math talk communities. *Teaching Children Mathematics, 22*(4), 248–254.

Wessman-Enzinger, N. M. (2018). Integer play and playing with integers. In L. Bofferding & N. M. Wessman-Enzinger (Eds.), *Exploring the integer addition and subtraction landscape: Perspectives on integer thinking* (pp. 25–46). Cham, Switzerland: Springer.

Wessman-Enzinger, N. M. (2019). Integers as directed quantities. In A. Norton & M. Alibali (Eds.), *Constructing number* (pp. 279–305). Cham, Switzerland: Springer.

Wessman-Enzinger, N. M. (2023). Build it! Imagining integers. *Mathematics Teacher: Learning and Teaching PK–12, 116*(10), 771–775.

Wessman-Enzinger, N. M., & Bofferding, L. (2023). Beyond the statistics: Joy in mathematics. In T. Lamberg & D. Moss (Eds.), *Proceedings of the forty-fifth annual meeting of the North American chapter of the International Group for the Psychology of Mathematics Education* (Vol. 2, pp. 26–34). Reno, Nevada: University of Nevada, Reno.

Wessman-Enzinger, N. M., & Gerstenschlager, N. E. (2023). Unpacking elementary preservice teachers' ways of reflecting on conceptual mistakes. *Investigations in Mathematics Learning, 15*(3), 186–204. https://doi.org/10.1080/19477503.2023.2187167

Wiles, A. (1995). Modular elliptic curves and Fermat's Last Theorem. *Annals of Mathematics, 141*(3), 443–551.

Willingham, D. T. (2009). Is it true that some people just can't do math? *American Educator*, 14–19.

Willingham, J. C., Strayer, J. F., Barlow, A. T., & Lischka, A. E. (2018). Examining mistakes to shift student thinking. *Mathematics Teaching in the Middle School, 23*(6), 324–332.

Yildiz, A. (2013). Views of pre-service teachers related to the mistakes encountered in the textbooks which they benefited from in physics courses. *International Journal of Academic Research, 5*(4), 406–411.

Index

A

Allen, K., 13

analogical reasoning, 31, 33, 128

anchor charts, 87, 112

anxiety, 12

asset-based perspectives

    laying the foundation for an asset-based perspective, 85–88

    mistakes and, 13–15, 180

attribution of mistakes, 52. *See also* unknown student work

B

beautiful and powerful mistakes

    beauty and power in mathematics, 22–25

    big idea for, 21, 177

    concluding remarks for, 34–35

    defining beauty in mathematics, 23–24

    defining power in mathematics, 24–25

    negative integers as creative space for mistake making, 25–34

    reflection from Nicole, 21–22

    reflection questions for, 35

    reproducibles for, 36

Boaler, J., 158

C

celebrating mathematical mistakes

    celebration boards, 16

    eliciting and celebrating mistakes, 71–73

    three main ideas for celebrating, 9

    views of mistakes and, 15–17

changing minds in mathematics

    about, 104

    big idea for, 103, 178

    changing minds as a creative act, 105–109

    changing minds as a pedagogical tool, 109–116

    changing minds as a task, 116–118

    changing minds in mathematics, 104–118

    concluding remarks for, 118–119

    layers of, 105

mistakes in action, 182, 183

reflection from Nicole, 103–104

reflection questions for, 119

reproducibles for, 120–121

classroom norms, 86–87, 163

closed-ended tasks, 95–96, 98, 99.
See also tasks

closed-middle tasks, 95, 96–97, 98.
See also tasks

complex instruction, use of term, 85

conceptual mistakes.
See also factual, procedural, and conceptual mistakes

about, 53–54

definition of conceptual knowledge and mistakes, 41

kindergarten example: addition and conceptual mistakes, 54–58

"My Favorite Conceptual Mistake" project, 73

conceptual understanding

celebrating mathematical mistakes and, 17

conceptual mistakes, 53–58, 73

deep mathematical discussions and, 114

factual mistakes and, 45

gameplay and, 163, 165

mathematical argumentation and justification and, 23

this or that tasks and, 124–125, 127–130

Warren's strategy and, 32

continuum of favorite mistakes.
See also mistakes by mathematicians

mistakes and community: the case of 0.999, 66–68

mistakes and society: the case of zero and negative numbers, 66

mistakes from research: Fermat's last theorem, 68–70

countering, 31–33

counting

from closed-middle to open-middle tasks, 98

different bases (the case of base 5), 144–147

creativity

changing minds as a creative act, 105–109

creative thinking, 160

negative integers as creative space for mistake making, 25–34

culturally responsive teaching, use of term, 85

D

deep mathematical discussions, 113, 114, 115

deficit-based perspectives

deficit noticing, 85

shifting views of mistakes, 13–15

dice, example of unknown student work in probabilities, 91–93

different bases (the case of base 5), 144–147, 148

E

Edison, T., 189

eliciting surprise

defining beauty in mathematics and, 23, 24

Gertrude's strategy, 28

rich tasks and, 71

Warren's strategy, 33

F

factual, procedural, and conceptual mistakes.
See also conceptual mistakes; factual mistakes; procedural mistakes

big idea for, 39, 177

concluding remarks for, 58

mistakes in action, 181

reflection from Natasha, 39–40

reflection questions for, 58–59

reproducibles for, 60

types of mistakes, 40–58

factual mistakes.
See also factual, procedural, and conceptual mistakes

about, 44–45

definition of factual knowledge and mistakes, 41

fifth-grade example: fractions and factual mistakes, 45–47

high school example: geometry and factual mistakes, 47–49

Featherstone, H., 158

Fermat's last theorem, 68–70

focusing questions, identifying, 166–167

funneling questions, 166

G

gallery walks, 113, 114

gameplay.
See also mathematical games

about, 161–163

Index

identifying focusing questions, 166–167
identifying mathematical goals, 163–164
identifying the game, 164–165
monitoring implementation and preparing for synthesis, 167
synthesizing learning, 168–169
tenets of play, 158
generating new ideas, 23, 27, 28, 29
Ginsburg, H., 157
growth mindset, 17

I

identities.
See mathematical identities
instructional strategies for examining mistakes
about, 84–85
asset-based perspectives, laying foundation for, 85–88
big idea for, 83, 178
concluding remarks for, 99
reflection from Natasha, 83–84
reflection questions for, 99
reproducibles for, 100–101
tasks, converting or revising existing, 97–99
tasks to promote the inspection-worthy mistakes strategy, selecting, 94–97
unknown student work strategy, using, 87–93
introduction, 1–5
invented notation and language
about, 137
big idea for, 135, 178
concluding remarks for, 149–150
different bases (the case of base 5), 144–147
mistakes in action, 184
negative numbers on the right-hand side, 140–141
reflection from Nicole, 135–137
reflection questions for, 150
reproducibles for, 151–153
saying numbers using *and*, 142–143
using and supporting in the classroom, 147–149
using invented notation and language, 137–147
zero numbers and, 137–139

K

keep-change-flip, 90

L

limitations in mathematics, 23, 27
linking cube representations, 28, 146
Long, J., 69, 70

M

mathematical games
about, 156
big idea for, 155, 178
concluding remarks for, 169
gameplay, 161–169
mathematical play, 156–161
reflection from Natasha, 155–156
reflection questions for, 169
reproducibles for, 170–172
mathematical identities
about, 13
celebrating mathematical mistakes and, 17, 72, 73
confidence and, 70
normalizing mistakes and, 65
student's drawing depicting self when the focus was on correctness alone, 74
mathematical mistakes.
See also beautiful and powerful mistakes; celebrating mathematical mistakes; factual, procedural, and conceptual mistakes; instructional strategies for examining mistakes; mistakes by mathematicians; mistakes in action; shifted view of mistakes
continuum of favorite mistakes, 65–70
identifying mistakes in practice, 42–43
types of mistakes, 40–58
use of term, 14–15
Mathematical Practice 1, 24
mistakes by mathematicians
about, 64–65
acts of celebrating, 73
big idea for, 63, 177
concluding remarks for, 74
continuum of favorite mistakes, 65–70
eliciting and celebrating mistakes, 71–73
reflection from Nicole, 63–64

reflection questions for, 74–75
reproducibles for, 76
mistakes in action
    big idea for, 175, 178
    big ideas from each chapter, revisiting, 177–185
    concluding remarks for, 185
    reflection from Natasha, 176
    reflection from Nicole, 176
    reflection questions for, 185
    reproducibles for, 186
"My Favorite Conceptual Mistake" project, 73

N

National Council of Teachers of Mathematics (NCTM), 110, 163
negative integers
    mistakes and society: the case of zero and negative numbers, 66
    negative numbers on the right-hand side, 140–141
    reflection from Nicole, 135–136
    tasks that support imagination of mathematical topics and, 148, 149
negative integers as creative space for mistake making
    about, 25–26
    better together with both correct and incorrect solutions, 29, 33
    Gertrude's strategy, 27–28
    Miguel's strategy, 30–31
    Taylor's strategy, 26–27
    Warren's strategy, 31–33
normalizing mistakes
    celebrating mathematical mistakes and, 16, 73
    mathematical identities and, 65
    mistakes and community and, 67–68
    this or that tasks and, 127
norms, 86–87, 163
number talks, 113–114

O

open-ended tasks, 25, 54, 96, 97, 98, 99. *See also* tasks
open-middle tasks, 54, 71, 95–96, 97, 98. *See also* tasks
order in mathematics, 23

P

pedagogical tools, changing mind as
    about, 109–110
    layers of changing minds, 105
    supporting, 113–114
    vignettes for, 110–112, 114–116
perimeter, shape for determining perimeter using the darker shaded part, 59
persevering, 109–110
pervasive mistakes
    conceptual mistakes, 55
    factual mistakes, 45, 49
    mistakes and community and, 67
    procedural mistakes, 49, 51
positive integers, 136, 138
power in mathematics, defining, 24–25. *See also* beautiful and powerful mistakes
procedural fluency, 111, 125, 162, 165
procedural mistakes. *See also* factual, procedural, and conceptual mistakes
    about, 49
    definition of procedural knowledge and mistakes, 41
    third-grade example: addition and procedural mistakes, 50–53
productive struggle, 16

Q

questions, funneling, 166
questions, identifying focusing, 166–167

R

reproducibles for
    chapter 1 application guide, 18
    chapter 2 application guide, 36
    chapter 3 application guide, 60
    chapter 4 application guide, 76
    chapter 5 application guide, 100
    chapter 5: two foundational instructional strategies for examining mistakes, 101
    chapter 6 application guide, 120
    chapter 6: changing minds task structure, 121
    chapter 7 application guide, 132
    chapter 7: this or that task structure, 133
    chapter 8 application guide, 151

chapter 8: invented notation and language, 152–153

chapter 9 application guide, 170

chapter 9: resources for mathematical game play, 171–172

chapter 10 application guide, 186

rich tasks, attributes of, 71.
See also tasks

S

saying numbers using *and*, 142–143.
See also invented notation and language

Schnell, K., 13

sense making, 24, 31, 33, 86

sharing mistakes, 52, 67.
See also unknown student work

shifted view of mistakes

    about, 12

    big idea for, 11, 177

    celebrating mathematical mistakes, 15–17

    concluding remarks for, 17

    reflection from Natasha, 11–12

    reflection questions for, 17

    reproducibles for, 18

    shifting views of mistakes, 12–15

Smith, M., 71

social engagement, 158, 160

Stein, M., 71

strategies.
See also instructional strategies for examining mistakes

    selecting tasks to promote the inspection-worthy mistakes strategy, 94–97

    strategy work, 14

    unknown student work strategy, 87–93

structure identification, 24

surprise.
See eliciting surprise

T

tasks.
See also this or that tasks

    changing minds as a task, 116–118

    comparing and contrasting tasks A and B, 94–95

    comparing and contrasting tasks C and D, 95–97

    converting or revising existing tasks, 97–99

    invented notation and language and, 148–149

    layers of changing minds, 105

    selecting tasks to promote the inspection-worthy mistakes strategy, 94–97

think-pair-shares, 113, 114, 115

this or that tasks.
See also tasks

    about, 124

    big idea for, 123, 178

    concluding remarks for, 130

    creating, 125–130

    mistakes in action, 182

    reflection from Nicole, 123–124

    reflection questions for, 131

    reproducibles for, 132–133

    this or that fraction comparison tasks, 126

    This or That Task Structure, 124–125

Tiny Polka Dot, 162, 164, 166

two foundational instructional strategies for examining mistakes.
See instructional strategies for examining mistakes

Tyson, N., 1

U

unknown student work

    example of unknown student work in probabilities, 91–93

    example of unknown student work in the division of fractions, 88–91

    use of term, 84

    using the unknown student work strategy, 87–93

W

Wiles, A., 109

Willingham, D., 44, 49, 53

Z

zero numbers and invented notation and language, 137–139

Common Denominators
Jennifer A. Lenhardt
"*Common Denominators* is a collection of stories braided together with research-informed strategies and tools," writes author Jennifer A. Lenhardt. Make sense of student engagement and belonging by using mathematics concepts that illustrate our common humanity and illuminate a clear, sustainable path for honoring and meeting all students' needs.
BKG179

Nurturing Math Curiosity With Learners in Grades K–2
Chepina Rumsey and Jody Guarino
Gain the educational tools needed for planning, communicating, and representing mathematical ideas to students. This book gives teachers instructional strategies to enhance their students' natural wonder and curiosity toward math concepts.
BKG180

See It, Say It, Symbolize It
Patrick L. Sullivan
Anyone can learn mathematics and stay in the math game for life once they learn key superpowers that can demystify foundational concepts—from whole numbers, fractions, and place value operations to ratios, proportions, and percentages. This book offers teaching methods to develop a dynamic and flexible understanding of numbers and operations in young learners.
BKG187

The SNAP Solution
Kirk Savage, Jonathan Ferris, and Tom Hierck
Discover student numeracy assessment and practice (SNAP)—a practical approach to help classroom teachers evaluate number sense. With its simple and reliable method, K–8 math teachers can quickly implement SNAP in their daily teaching practices and ignite a sense of wonder and thinking for math in their students.
BKG199

Visit SolutionTree.com or call 800.733.6786 to order.

Global PD teams
Collaborative Learning for School Improvement

Quality team learning **from authors you trust**

Global PD Teams is the first-ever **online professional development resource designed to support your entire faculty on your learning journey.** This convenient tool offers daily access to videos, mini-courses, eBooks, articles, and more packed with insights and research-backed strategies you can use immediately.

GET STARTED
SolutionTree.com/**GlobalPDTeams**
800.733.6786